Foundations of Surface Science

Foundations of Surface Science

Stephen J. Jenkins

Professor of Physical and Computational Surface Chemistry

Yusuf Hamied Department of Chemistry, University of Cambridge

OXFORD

UNIVERSITY PRESS

Great Clarendon Street, Oxford, OX2 6DP,
United Kingdom

Oxford University Press is a department of the University of Oxford.
It furthers the University's objective of excellence in research, scholarship,
and education by publishing worldwide. Oxford is a registered trade mark of
Oxford University Press in the UK and in certain other countries

Published in the United States of America by Oxford University Press
198 Madison Avenue, New York, NY 10016, United States of America

British Library Cataloguing in Publication Data
Data available

Library of Congress Control Number: 2022944958

ISBN 978-0-19-883513-4

Printed and bound by CPI Group (UK) Ltd, Croydon, CR0 4YY

Preface

Errors, like straws, upon the surface flow;

He who would search for pearls must dive below.

John Dryden

It has long been understood that surfaces are special. No matter how perfect a material might be, the abrupt termination of its ideal structure and symmetry implies not only a reduction from three dimensions to two, but also the inevitability of unique sites and defects unknown in the underlying bulk. If we are searching for perfection, the surface is not the place to find it. But if we are looking for interesting and exploitable phenomena, it is a great place to start. Conversely, if our favourite bulk material fails to live up to expectations in practice, it may be that the properties of its surface are to blame. To understand these properties at an atomic scale, to avoid or to exploit them as the need or opportunity arises, is thus the proper goal of surface science.

From the chemist's perspective, two historical applications of this approach loom particularly large, namely catalysis and corrosion. Common rust, to take just one example of the latter, is formed by the action of oxygen and water at iron surfaces, and although the result is readily visible to the naked eye, we must not forget that what we see is the cumulative product of countless individual chemical reactions taking place at the nanoscale. Modification of the surface, perhaps only one or two atomic layers deep, can therefore have a disproportionate effect in slowing the progress of corrosion. If we can grasp the fundamentals, we stand a chance of proposing new surface treatments, prolonging the lifetime of metal components across a variety of environments. In heterogeneous catalysis, on the other hand, we are interested in speeding up preferred reactions, to enhance either activity or selectivity, but once again it is what occurs at the surface that critically dictates our success or failure. Here, it is crucial that we understand the nature of catalytically active surface sites, and the role that surface modifiers might play in promoting or poisoning their influence.

More recently, however, the importance of surface science across other fields has become increasingly apparent. We might mention, for example, the rise of microfluidics, in which surface tension may dominate bulk properties in determining flow dynamics and mixing. In the solid state, on the other hand, we could speak of nanoparticles, whose shapes are determined by surface energy, and whose reactivity may be dominated by high-energy edge and corner sites. In electronics, the importance of surface defects rises dramatically as components are progressively miniaturised, to the extent that doping by surface treatment becomes viable when bulk doping fails. Indeed, the very fabrication of semiconductor devices, by means of chemical vapour deposition, relies heavily on understanding the chemistry of volatile precursors as they settle upon, react with, and incorporate into a growing solid surface. And the use of functionalised surfaces for chemical and biological sensing applications builds upon venerable principles of molecular recognition, given new life by advances in surface measurement techniques.

This book does not, however, focus upon any of the exciting applications mentioned above, for the simple reason that it would likely be out-of-date within months, if not before publication. Instead, it aims to provide an overview of fundamental surface science, which the reader can then apply to whatever applications they wish. To this end, the contents are arranged thematically, according to quite traditional principles. First, a chapter on surface thermodynamics addresses foundational issues, including where the surface is actually located, and what meaning we can attach to terms such as surface energy, surface tension, and surface stress; here we deal also with atoms or molecules that bind to surfaces, whether described as surfactants (at liquid surfaces) or adsorbates (at solid surfaces). Next, a chapter on surface symmetry and structure provides concepts and terminology to understand how real surfaces may differ from the ideal truncation of bulk crystallinity, whether by relaxation, reconstruction, or adsorption. A third chapter, covering electronic structure, examines the underlying driving forces behind physisorption and chemisorption, alongside discussion of such key concepts as the surface dipole and the work function; surface-localised electronic states, and their consequences, are also examined in some depth. Finally, so far as our thematic treatment is concerned, a chapter on kinetics and dynamics addresses the processes that occur at surfaces, whether relating to adsorption, desorption, vibration, or chemical reaction.

Tying all the preceding chapters together, the book ends with a compendium of techniques commonly applied to surfaces. It is important to stress that surface science, perhaps more than any other field of study, is a discipline in which it is rare for a problem to be solved by the application of just one technique. In recent years, for example, scanning probe microscopy has proven to be an extraordinarily useful tool, granting access to real-space images of surfaces at a resolution that would have been unthinkable just a few decades ago. Nevertheless, without insight from many of the more traditional surface science techniques, interpretation of scanning probe images would rest on rather shaky ground. Likewise, the burgeoning role played by computational surface studies has undoubtedly been massively beneficial to the field, but as a practitioner the author feels obliged to point out its limitations and to advise appropriate caution. Theory is often at its best when deployed alongside experiment, and it is incumbent upon both experimentalists and theorists to understand each other's methods. For this reason, the book ends by briefly summarising the application of density functional theory to surface systems.

Before any of that, however, I would like to take this opportunity to thank those without whom there would be no need for this preface, starting with Prof. G. P. Srivastava and Prof. Sir David King, who each taught me more about surfaces than books ever could. In addition to several close collaborators and colleagues (most notably, Prof. Georg Held, Prof. David Wales, and Dr Marco Sacchi), I am also indebted to all members of my research group, past and present, whether mentioned here by name or not.

Particular thanks are due to Prof. Stuart Clarke, Dr Stephen Driver, Dr Vittorio Fiorin, Dr David Madden, and Dr Israel Temprano, who have all, at various times, been responsible for developing and/or teaching the third-year course on Surfaces and Interfaces, in the Chemistry Department at Cambridge, which heavily inspired parts of the present book. I am grateful also to the students who have taken that course over the years, not least for their ability to spot errors in lectures and notes, and also to those who have attended my fourth-year course on the Electronic Structure of Solid Surfaces, for the same reason. Needless to say, remaining errors and omissions in the present text are my responsibility alone, but special thanks are due to Prof. Clarke and Dr Madden for their close reading of early drafts, not to mention their encouragement along the way. And finally, heartfelt thanks to Victoria, my wife, for supporting me in writing a second book, despite my having promised never to write one again.

Stephen J. Jenkins
Cambridge, 28 February 2022

Contents

Preface   v

1 Thermodynamics   1

1.1   Introduction   1

1.2   Dividing plane and surface excess   1

1.3   Specific surface energy and surface free energy   4

1.4   Surface tension and surface stress   5

1.5   Surface curvature and its consequences   11

1.6   Surfactants and the Gibbs isotherm   14

1.7   Gas/solid isotherms and relative coverage   17

1.8   Heats of adsorption and lateral interactions   22

1.9   Exercises   25

1.10 Summary   25

     Further reading   26

2 Symmetry and Structure   27

2.1   Introduction   27

2.2   Bulk lattices and crystals   27

2.3   Miller indices and ideal surfaces   30

2.4   Relaxation and reconstruction   37

2.5   Notation for superstructure   40

2.6   Reciprocal space   42

2.7   Exercises   45

2.8   Summary   46

     Further reading   47

3 Electronic Structure   48

3.1   Introduction   48

3.2   Surface dipole and work function   48

3.3   Surface-localised electronic states   56

3.4   Adsorbate-surface bonding   64

3.5   Exercises   68

3.6   Summary   69

     Further reading   70

4 Kinetics and Dynamics   71

    4.1   Introduction   71
    4.2   Adsorption   71
    4.3   Desorption   78
    4.4   Vibration   82
    4.5   Reaction   87
    4.6   Exercises   91
    4.7   Summary   91
          Further reading   92

5 Techniques   94

    5.1   Introduction   94
    5.2   Electron diffraction techniques   95
    5.3   Scanning probe techniques   97
    5.4   Photoemission techniques   102
    5.5   Auger Electron Spectroscopy (AES)   105
    5.6   Near-Edge X-ray Absorption Fine Structure (NEXAFS)   107
    5.7   Molecular beam techniques   108
    5.8   Vibrational spectroscopies   112
    5.9   Density Functional Theory (DFT)   115
    5.10 Exercises   120
    5.11 Summary   121
          Further reading   121

Glossary   123
Index   128

Thermodynamics

1.1 Introduction

Before tackling surface thermodynamics, it is sensible to begin by asking just what we actually mean when we refer to the 'surface' of a substance. Once this rather crucial point is clarified, much of the necessary conceptual framework turns out to be closely analogous to our pre-existing understanding of bulk thermodynamics. Guided by this analogy, we shall proceed by first exploring the concept of surface free energy—that property whose minimisation may be regarded as the prime mover of all surface processes—and the related concepts of surface tension (liquid surfaces) and surface stress (solid surfaces). From this foundation we shall explore the consequences of surface curvature, before considering the role of surfactants in modifying surface tension. Finally, we shall address adsorption, discussing a number of common gas/solid isotherms and assessing how the heat of adsorption may be affected by interadsorbate lateral interactions.

1.2 Dividing plane and surface excess

As mentioned previously, our most pressing concern at the outset must be the definition of the surface itself. Let us begin, therefore, by considering a condensed medium (liquid or solid) comprising only a single chemical species, focussing specifically upon the particle density profile in the vicinity of what we shall tentatively identify as its surface.[1] Far beneath the surface, it is reasonable to assume that the particle density maintains a near-constant value, essentially identical to that of the bulk material. Close to the surface, however, one expects that the particle density must fall to zero while passing into the space beyond. It is reasonable to suppose that the transition from one constant-density regime (bulk) to the other (vacuum) is likely to be fairly abrupt but not necessarily perfectly so, leading to ambiguity in deciding where precisely the bulk region ends and the vacuum region begins.

---

[1] We shall, without loss of generality, consider the particle density (expressed as the number of moles per unit volume, whether of atoms or of molecules) as a continuum property.

A full thermodynamic treatment of this situation would appear, therefore, to require that we account for a somewhat gradual change of properties in passing through this boundary region, with all the attendant difficulties such a task would entail. Instead, however, it is possible to make progress within a much simpler representation of the system, by the naive expedient of *pretending* that the particle density does, in fact, collapse abruptly from its bulk value to zero at some particular plane (the **dividing plane**) that we shall identify with the division between inside and outside the surface. In doing so, we naturally make an error, and it is this error that we must account for (and feed back into our model) through a property known as the **surface excess**.

Specifically, consider the situation depicted in Fig. 1.1a, where the dividing plane has been situated at a location where the particle density has declined only slightly relative to its bulk value. Our model now consists of three regions: (I) a region with constant bulk-like particle density throughout the entire space below the dividing plane; (II) a region with zero particle density throughout the entire space above the dividing plane; and (III) a notional region of zero thickness, situated at the dividing plane itself, encompassing the net missing material from the first two regions. The material contained within Region I of our model will be an overestimate relative to the real system, because our model neglects any decline in particle density as one approaches the dividing plane from below. The material contained within Region II will, in contrast, represent an underestimate relative to the real system, because our model neglects any particle density found above the dividing plane. As depicted, the result is that some balance remains unaccounted for between these two regions, and this must be deposited under the title of Region III. It is this material (quantified in moles per unit area) that is referred to as the surface excess, and we shall give it the symbol Γ.

It is worth emphasising that our choice of location for the dividing plane was essentially arbitrary. Fig. 1.1c shows the same system with the dividing plane situated at a position where the particle density has fallen rather close to zero. Once again, we partition our model into three regions, but now the material missing from Region II (zero particle density) is more than compensated by the material spuriously included in Region I (bulk-like particle density) and so the surface excess attributed to Region III (the dividing plane itself) must actually be negative. Evidently, it is always possible, in the case of a single-species system, to carefully position the dividing plane in such a way that the surface excess vanishes (Fig. 1.1b).

Matters become more complicated when multiple chemical species exist within the system, each with their own particle density profile in the vicinity of the surface. As before, we define the surface excess (for any given species) as the quantity of material (in moles per unit area) unaccounted for by regions of constant bulk-like particle density below the dividing plane (Region I) and zero particle density above it (Region II). We must therefore recognise multiple surface excesses, denoted by Γ_i, where the subscript labels the species. It will not, however, always be possible to locate the dividing plane in such a way as to ensure that the surface excess of all species will vanish. Nevertheless, in cases where the abundance of one species clearly dominates all others, the dividing plane

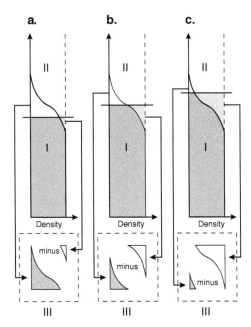

Fig. 1.1 Dividing planes and surface excess.

might most sensibly be chosen such that the surface excess of this single species vanishes, conceding that the surface excesses of all other species may remain non-zero (Fig. 1.2). Such a choice may be convenient, for example, when solute species are dissolved at low concentration within a single-component solvent.

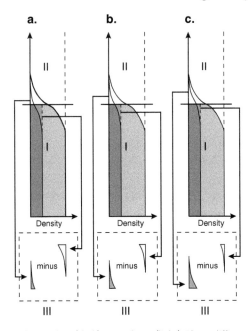

Fig. 1.2 Surface excess for a solute (dark) in a solvent (light). Three different density profiles are shown, all with the same dividing plane.

1.3 Specific surface energy and surface free energy

The **internal energy** of a thermodynamic system includes all the kinetic
and potential energies of its constituent
particles, but not the kinetic or potential
energy due to the wholesale motion or
position of the system itself.

Readers familiar with bulk thermodynamics will recall that the concept of **internal energy**, conventionally denoted by the symbol U, is of central importance to the topic. It is defined as the sum of all kinetic and potential energies attributable to the particles within a system (atoms and/or molecules) excluding any kinetic energy associated with the system's wholesale motion and any potential energy due to the system's overall position within an external field. Infinitesimal changes in the internal energy may then be expressed according to the master equation

$$dU = TdS + dW + \sum_i \mu_i \, dN_i \qquad [1.1]$$

where T represents the system's temperature and S its entropy. The symbols μ_i then represent the molar chemical potentials of the system's constituent species, while the corresponding quantities of particles (in moles) are denoted by N_i. Under reversible conditions, the first term represents heat flow into or out of the system, while the third term accounts for chemical energy carried into or out of the system by exchange of particles with its surroundings. The second term, dW, corresponds to reversible work done upon infinitesimal deformation of the system (changing either its size or its shape), and its precise form may differ depending upon whether we are dealing with a solid or a fluid system.[2]

Now, the internal energy is not always the most convenient measure of the system's capacity (or 'potential') to perform work under different circumstances, and for this reason a variety of alternative **thermodynamic potentials** have historically been introduced, including the **Helmholtz energy**, F, defined by

Thermodynamic potentials are
descriptors that quantify the ability of a
system to do reversible work under certain specified conditions (e.g. constant
pressure).

$$F = U - TS \qquad [1.2]$$

and the **grand potential**, Φ, defined by

The Helmholtz energy of a system
quantifies its ability to do reversible
work under conditions of constant
temperature and constant particle
quantities.

$$\Phi = F - \sum_i \mu_i N_i \qquad [1.3]$$

The **grand potential** of a system quantifies its ability to do reversible work under conditions of constant temperature
and constant chemical potentials.

to address precisely this issue. Specifically, we shall state (without proof) that F represents the system's capacity to do reversible work under conditions of constant temperature and particle quantities (i.e. $dT = dN_i = 0$), while Φ represents the system's capacity to do reversible work under conditions of constant temperature and chemical potentials (i.e. $dT = d\mu_i = 0$). For this reason, both are considered to be types of 'free' energy, where the word 'free' is used in the sense of 'available (under certain conditions)'. Both will be useful to us in due course.

Surface excess internal energy
accounts for the internal energy associated with a system's surface excess,
supplemented by the reversible work of
surface creation. The same logic applies
also to the surface excess versions of
other thermodynamic potentials (e.g.
surface excess Helmholtz energy
and **surface excess grand potential**).

The keystone of surface thermodynamics is the notion of **surface excess internal energy**, which we should understand to mean the internal energy

[2] It is worth noting that Eqn. 1.1 holds under all conditions, but the neat interpretation of its terms presented here is valid only for processes that are reversible. In practice, a reversible process is one that takes place infinitesimally slowly (quasistatically).

associated with Region III in the model introduced previously. We shall follow Gibbs (1877) in writing this as

$$U^s = TS^s + \gamma A + \sum_i \mu_i N_i^s \qquad [1.4]$$

where the superscript 's' should be taken to imply a surface excess quantity (belonging exclusively to Region III). The symbol S^s, for example, refers to the **surface excess entropy**, while the quantity of particles associated with the surface, N_i^s, is synonymous with $\Gamma_i A$, where A represents the surface area and Γ_i was defined previously. The second term, γA, accounts for reversible work done in breaking whatever bonds (physical or chemical) that needed to be broken to bring the surface into existence in the first place, with γ being the reversible work done per unit area (or **specific surface energy**). If, for instance, a bulk sample were cleaved to create two identical fresh surfaces of total area $2A$ (see Fig. 1.3), then we would consider the reversible work of surface creation (γA) for each of those surfaces to be half of the total reversible work involved.[3]

As with the bulk case, however, internal energy is not necessarily the most useful thermodynamic potential for our purposes, and once again it is often better to focus upon one or other of the free energies. Notably, although we may sometimes wish to discuss the **surface excess Helmholtz energy**, F^s, defined as

$$F^s = U^s - TS^s$$
$$= \gamma A + \sum_i \mu_i N_i^s \qquad [1.5]$$

it will often be particularly useful to concentrate on the **surface excess grand potential**, Φ^s, which takes the form

$$\Phi^s = F^s - \sum_i \mu_i N_i^s$$
$$= \gamma A \qquad [1.6]$$

since it is explicitly independent of the position chosen for the dividing plane. Indeed, Φ^s is usally regarded as 'the' **surface free energy**, and it is Φ^s that the surface seeks to minimise at equilibrium.[4]

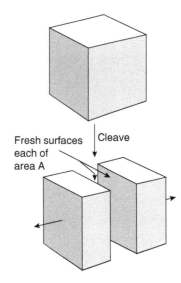

Fig. 1.3 Creation of surfaces.

Surface excess entropy refers to the entropy associated with the system's surface excess material.

Specific surface energy refers to the reversible work done in creating fresh surfaces by cleaving a bulk sample, normalised to the total area of surface thus created.

Surface free energy is a rather loose term, but usually refers to the surface excess grand potential.

1.4 Surface tension and surface stress

Having just defined one key property of our system, it is a natural scientific instinct to ask ourselves how it may vary under an infinitesimal perturbation. How, in this case, might we expect the surface free energy to change in response to a gradual deformation of the surface, and what might that imply about the surface's resistance to such deformation?

---

[3] If the process of cleavage produces two dissimilar surfaces, or if the surface stoichiometry differs from that of the bulk, it is clear that a more nuanced picture of surface creation will be required, but there will still be a definite amount of reversible work involved and it will still be proportional to the surface area.

[4] Note how Eqns. 1.5 and 1.6 (relating to surface excess potentials) share the same structure as Eqns. 1.2 and 1.3 (relating to bulk potentials).

Surface tension of a liquid

Imagine a quantity of liquid, contained within a rectangular trough, whose surface is divided in two by a mobile floating boom (Fig. 1.4).[5] For thermodynamic purposes we shall consider the liquid surface on one side of the boom to comprise the whole of our system, whilst that on the other side, together with the bulk of the liquid and the apparatus itself, will be treated as part of the surroundings.

Next, imagine that the boom is very slowly moved, so that the system's area, A, is expanded by an infinitesimal increment, dA. Clearly, a certain amount of work may be done on the system in the process, and if we wish to quantify this it will be necessary to examine changes in the surface excess grand potential, Φ^s, which we have previously identified (Section 1.3) as the system's capacity to do reversible work under conditions of constant temperature and chemical potentials. The chemical potentials in our current system are, of course, fixed through contact with an external reservoir (i.e. the bulk). We may therefore write

$$dW^s = d\Phi^s = d(\gamma A)$$
$$= \gamma\,dA + A\,d\gamma \qquad\qquad [1.7]$$

for the reversible work done on the surface.

We can, however, go a little further, by recognising that the specific surface energy, γ, cannot vary when dealing with a liquid surface. Whether the surface expands or contracts, particles will migrate from or to the bulk liquid, ensuring that the local arrangement of particles at the surface (upon which γ solely depends) remains unchanged. The second term in Eqn. 1.7 thus vanishes, and we are free to write

$$dW^s = \gamma\,dA$$
$$= \gamma y\,dx \qquad\qquad [1.8]$$

where we have introduced y as the width of the trough, and dx as the distance through which the boom has moved. Moreover, rearranging this into the form

$$\gamma = \frac{1}{y}\frac{dW^s}{dx} \qquad\qquad [1.9]$$

we see that the specific surface energy (γ) is synonymous with the force (dW^s/dx) per unit width (y) against which the mobile boom must work in expanding the surface. That is to say, the reversible work done in expanding the surface is precisely equivalent to that which would be expected if the liquid surface were replaced by a thin elastic sheet of tension equal to γ. For this reason, the specific surface

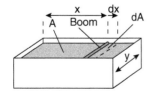

Fig. 1.4 Trough for surface tension experiments.

[5] Troughs of similar design have been employed for many years in the study of liquid surfaces and layers of molecules deposited thereon. Developed and popularised by Irving Langmuir and Katharine Blodgett (and named Langmuir–Blodgett troughs, in their honour) all such troughs ultimately derive, in fact, from the one first described by Agnes Pockels in a letter to Lord Rayleigh (see Rayleigh, 1891).

energy of a liquid is often referred to as its **surface tension**. It would be as well to remember, however, that the two terms technically refer to different properties, albeit ones that happen to be dimensionally equivalent ('energy per unit area' being identical with 'force per unit width') and to share precisely the same value.

Surface tension is that property of a liquid surface whereby it tends to resist expansion of the surface area. Expressed as a force per unit length, its numerical value is equal to the specific surface energy (usually stated as an energy per unit area).

Surface stress of a solid

In considering the variation of surface free energy upon deformation of a solid surface, we are clearly not able to envisage varying its size or shape independently of the underlying bulk, as was possible above in the case of a liquid surface. Instead, we must imagine the entire solid to be somehow clamped into a vice capable of imposing a well-defined deformation (strain) on demand, accepting that both bulk and surface will be subject to the same distortion. Since we are concerned here only with surface phenomena, however, we shall at least restrict ourselves to deformations entirely within the plane of the surface (**surface strain**).

Surface strain is a tensor that quantifies the deformation of a solid surface, either in a sense that alters the surface area (**normal strain**) or in a sense that does not (**shear strain**).

To provide a convenient reference frame, imagine that we have painted a small square onto the surface of our sample, with sides aligned along arbitrary axes $\hat{\mathbf{x}}_1$ and $\hat{\mathbf{x}}_2$. Any possible uniform deformation applied to the surface will distort this square, but the resulting shape will retain parallel opposing sides. Furthermore, the surface strain may be resolved into a linear combination of four fundamental strain components, depicted in Fig. 1.5. These components are denoted symbolically as ε_{ij}, where each should be taken as indicating relative displacement in the $\hat{\mathbf{x}}_j$ direction of those sides lying perpendicular to the $\hat{\mathbf{x}}_i$ direction, with magnitude expressed as a dimensionless fraction of the original square's side length, L. That is, for example, a strain expressed as $\varepsilon_{11} = 0.01$ implies that the square has been expanded in the $\hat{\mathbf{x}}_1$ direction by 1%. A strain expressed as $\varepsilon_{12} = 0.02$, on the other hand, would correspond to a distortion in which the sides perpendicular to the $\hat{\mathbf{x}}_1$ direction move apart in the $\hat{\mathbf{x}}_2$ direction by a relative displacement equal to 2% of the original side length. Strain components that alter the area of the square (ε_{11} and ε_{22}) are described as **normal strain**, while those that do not (ε_{12} and ε_{21}) constitute the **shear strain**.

Unlike expansion of a liquid surface, deformation of a solid surface does *not* take place with constant chemical potentials, as these will vary in response to the inevitable deformation of the underlying bulk. That said, infinitesimal strain of a solid surface ought not to cause any change in the total quantity of particles associated with that surface, meaning that the reversible work associated with infinitesimal deformation of the surface (at constant temperature) will be identical to the change in surface excess Helmholtz energy, dF^s. Thus, we may write

$$dW^s = dF^s = d(\gamma'A)$$
$$= \gamma'dA + Ad\gamma'$$

[1.10]

where γ' must be related to the specific surface energy, γ, via

$$\gamma' = \gamma + \sum_i \mu_i \Gamma_i$$

[1.11]

to ensure consistency with Eqn. 1.5.

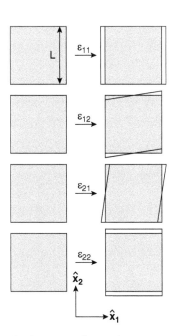

Fig. 1.5 Four surface strains.

Recognising that for solid surfaces the quantity γ' is explicitly dependent upon the strain we have imposed $(d\gamma' \neq 0)$ we must have (see Exercise 1.1)

$$dW^s = A\sum_{ij}\sigma_{ij}\,d\varepsilon_{ij} \qquad [1.12]$$

where σ_{ij} is the so-called **surface stress**, given by

$$\sigma_{ij} = \gamma'\delta_{ij} + \frac{\partial\gamma'}{\partial\varepsilon_{ij}} \qquad [1.13]$$

Surface stress is a tensor that quantifies the forces tending to oppose surface strain. These include both **normal stress** and **shear stress** components.

The **Shuttleworth equation** links surface stress to the changes in specific surface energy in response to surface strain.

which is known as the **Shuttleworth equation** (Shuttleworth, 1950). The concept of surface stress generalises that of surface tension, having not only non-zero **normal stress** (σ_{11} and σ_{22}) coupling to the normal strain components, but now also non-zero **shear stress** (σ_{12} and σ_{21}) coupling to the shear strain components.[6]

It is worth noting that the surface stress does, in general, depend upon our choice of dividing plane. Quite simply, the *total* stress within our system comprises not only the surface stress but also the stress inherent in the underlying bulk. Moving the dividing plane will alter the balance between these two contributions, so one ought strictly to specify its location when stating the corresponding surface stress. An important exception occurs, however, when the underlying bulk conforms to its equilibrium lattice constant, since in this case the bulk stress must be zero and the surface stress will be independent of the choice made for the dividing plane.[7]

At this point, one might reasonably ask what difference it may have made had we imagined our reference square in a different orientation. The components of surface strain would be differently defined, and correspondingly the components of surface stress would also take on different values. Nonetheless, their summed products (as per Eqn. 1.12) must be the same for any specific deformation of the surface, since our choice of reference orientation must be essentially arbitrary. That this holds true may be attributed to the fact that both surface strain, ε_{ij}, and surface stress, σ_{ij}, are tensor properties. This means that whilst their individual components may change when we rotate our coordinate system, they do so in such a way that the quantity they collectively express remains unaltered.[8]

One consequence of the tensor nature of surface stress is that it must always be possible to find a coordinate system for which the shear stress components

---

[6] Note that symmetry constrains the two shear stress components to be equal at equilibrium, in order that there be no unbalanced torque on the surface.

[7] If the surface conforms to its equilibrium lattice constant *and* has the same stoichiometry as the bulk, then it is actually permissible to substitute γ into the Shuttleworth equation, in place of γ' (see Exercise 1.1).

[8] The concept of a tensor is just an extension of the concept of a vector. A vector in two-dimensional space comprises two vector components, both of which change their value upon a rotation of the coordinate system whilst leaving the essence of the vector (its magnitude and direction) unaltered. A tensor in two-dimensional space comprises four tensor components, all of which change their value upon rotation of the coordinate system whilst leaving the essence of the tensor (higher-order analogues of magnitude and direction) unaltered.

vanish. This can be achieved, in practice, by evaluating the four stress components within an arbitrary coordinate system, and then diagonalising the matrix

$$\begin{pmatrix} \sigma_{11} & \sigma_{12} \\ \sigma_{21} & \sigma_{22} \end{pmatrix}$$
[1.14]

whereupon the eigenvectors will be found to lie along the so-called **principal stress axes**, and the corresponding eigenvalues will be the so-called **principal stress components**. In general, the direction of the principal stress axes will not be known prior to careful measurement and/or detailed calculation (apart from the fact that they must be orthogonal to one another), and neither will the magnitudes of the principal stress components (which may not necessarily be equal to one another). Where a specific surface possesses particular symmetry elements, however, it is inevitable that the surface stress must respect these, so for certain high-symmetry cases the principal stress axes and/or principal stress components may be highly constrained.

Components of the surface stress tensor will vary when evaluated with respect to different arbitrary reference axes. For just one special set of axes, known as the **principal stress axes**, the shear stress components vanish. The remaining normal stress components, evaluated with respect to these special axes, are known as the **principal stress components**.

Measuring surface tension

Our treatment of surface tension and surface stress has thus far been somewhat abstract and theoretical, so it is high time to ask how we might actually measure these rather fundamental properties in practice. The boom-equipped trough from Fig. 1.4 might appear to permit measurement of a liquid's surface tension—by recording the force required to displace the boom—but a little thought quickly confirms that it really only grants access to the *difference* in surface tension on either side of the boom. We can thus measure changes in surface tension caused by somehow modifying the surface on just one side of the boom—dropping a little oil or soap onto it perhaps—but the absolute magnitude of the surface tension remains stubbornly inaccessible by such means. Shortly, we shall describe a classic alternative method that makes use of a dipping geometry to avoid this problem, but first we must go through some preliminaries. Specifically, we need to consider the wetting behaviour of liquids on solid surfaces.

When a liquid droplet sits upon a solid surface, the angle formed between the liquid surface and the underlying surface is known as the **contact angle**. In the case of **complete wetting**, the contact angle will be zero, whereas **complete dewetting** implies a contact angle of 180°.

Imagine, therefore, a droplet of liquid adhering to a solid in the geometry shown in Fig. 1.6, where the equilibrium **contact angle** is of prime importance. A contact angle approaching 0° represents the situation described as **complete wetting**, in which the droplet would spread to cover the entire solid surface, assuming sufficient liquid is present. A contact angle approaching 180°, in contrast, would represent **complete dewetting**, in which the droplet would form a compact ball in order to minimise its contact with the solid surface. More generally, the contact angle lies between these extremes, implying the condition of partial wetting. Note, however, that there is typically a small kinetic hysteresis in the value of contact angle obtained when studying a droplet that has attained its current size by the addition of liquid—the **advancing contact angle**—compared with a droplet that does so by the removal of liquid—the **receding contact angle**—with the equilibrium contact angle lying somewhere between these limits.

The measured contact angle of a droplet may depend upon whether it is increasing in size (**advancing contact angle**) or decreasing in size (**receding contact angle**). These may be viewed as bounding values for the equilibrium case.

We shall simplify our quantitative analysis considerably by assuming that the surface stress of the solid substrate is isotropic (as would be true either for

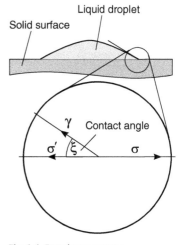

Fig. 1.6 Droplet geometry.

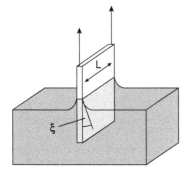

Fig. 1.7 Geometry of the Wilhelmy plate.

Young's equation relates the surface tension and contact angle of a liquid droplet with the surface stress of the material on which it rests, subject to an interfacial stress correction.

The cosine of contact angle for multiple liquids of differing surface tension on the same solid surface may usefully be presented in a **Zisman plot**, in order to determine the **critical surface tension** (at which the contact angle just vanishes). This quantity may hint towards the surface stress of the solid, but care must be exercised in the interpretation.

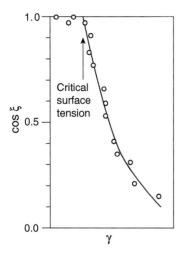

Fig. 1.8 Zisman plot. Data points are for different liquids on the same solid surface material.

a sufficiently high-symmetry single crystal or for an amorphous material) and hence representable as a scalar, σ. Where the solid surface is covered by the liquid, the interfacial stress would then remain isotropic but with reduced magnitude, σ'. Resolving lateral forces acting at the point of contact, we obtain

$$\gamma \cos \xi = \sigma - \sigma' \qquad [1.15]$$

where γ is the surface tension of the liquid and ξ the contact angle. This expression is known as **Young's equation** and works equally well when the solid is replaced with a liquid immiscible with the droplet (whereupon surface and interfacial stresses may be replaced with surface and interfacial tensions).

Having introduced the contact angle concept, we may now consider the experimental apparatus known as the Wilhelmy plate, illustrated in Fig. 1.7. Here, the weight of a thin solid plate is measured before and after partial immersion in the liquid of interest. Typically, the liquid partially wets the plate, forming a meniscus with a contact angle that may be measured visually (and which is often, in practice, very close to zero). The total downward force exerted on the plate by the surface tension of the liquid thus amounts to $2L\gamma \cos \xi$, where L is the lateral dimension of the plate and we have neglected forces acting upon its thin dimension. Since both L and ξ are known, the surface tension γ may be readily deduced simply by observing the change in apparent weight of the plate upon partial immersion. It must, of course, either be assumed that buoyancy effects are negligible, or a suitable correction must be included in the analysis.

Measuring surface stress

As hinted at above, the measurement of surface stress for a solid material is considerably more problematic than the measurement of surface tension for a liquid, even in the simplest isotropic case (Ibach, 1997). One approach that is commonly promoted in this regard involves measuring the equilibrium contact angle at the solid surface for liquids of differing known surface tensions. Graphing the cosine of contact angle versus the liquid surface tension, one obtains a so-called **Zisman plot** (Zisman, 1964), an example of which is presented in Fig. 1.8.

If a somewhat homologous series of liquids is used in construction of such a plot, then it is often possible to fit a reasonably smooth curve to the data for non-zero contact angles, and hence to determine the **critical surface tension** at which the curve intersects with unit contact-angle cosine. Examination of Young's equation (Eqn. 1.15) reveals that this critical value equates to the difference between the surface stress of the bare solid and the interfacial stress pertaining where the solid and liquid are in contact with one another. Since the latter quantity is, in general, both finite and unknown, it follows that the critical surface tension is emphatically *not* necessarily a measure of the solid's surface stress per se, although it is sometimes erroneously described as such in the literature. Indeed, a different homologous series of liquids may well yield a different critical surface tension for the same solid surface, and only if one were to find several such series converging upon the *same* critical surface tension might one tentatively speculate about some greater significance.

1.5 Surface curvature and its consequences

In our discussion of liquid droplets and menisci above, our analysis was based upon considering the liquid surface to be locally flat. A number of interesting phenomena arise, however, when we explicitly allow for the effects of global surface curvature.

Pressure differential across a curved liquid surface

Consider a spherical liquid droplet of volume V and surface area A, surrounded by its vapour. In the event of an infinitesimal virtual increase in the droplet's radius, r, the virtual work done, dW_v, is simply

$$dW_v = dW^s - \Delta P dV$$
$$= \gamma dA - \Delta P dV$$

[1.16]

where ΔP is the pressure differential between the inside and the outside of the droplet (positive values implying higher pressure within) and γ is the specific surface energy. We might profitably note, however, that V and A are not independent of one another, since both may be written parametrically in terms of the droplet radius. That is, we have $V = 4\pi r^3/3$ and $A = 4\pi r^2$, leading to derivatives $dV = 4\pi r^2 dr$ and $dA = 8\pi r dr$, allowing us to write

$$dW_v = 4\pi r(2\gamma - r\Delta P)dr$$

[1.17]

and hence to assert that at equilibrium (where virtual work necessarily vanishes for all virtual displacements) we must satisfy the condition

$$\frac{2\gamma}{r} = \Delta P$$

[1.18]

which is known as the **Young–Laplace equation**. Since γ is invariably positive, it follows that ΔP is also positive, implying that the pressure inside the droplet exceeds that outside. Note, importantly, that we could easily invert our physical situation—describing a vapour-filled bubble within a surrounding liquid environment—without changing any of the discussion just rehearsed. Whether we describe a liquid-filled droplet or a vapour-filled bubble, the pressure is always higher within the concavity of the surface, irrespective of which phase lies inside or outside of that boundary.

The **Young–Laplace equation** links the surface tension and radius of a curved surface with the pressure differential between its concave (high pressure) and convex (low pressure) sides.

Capillary action

One readily observable consequence of the pressure differential associated with curved liquid surfaces is the phenomenon of capillary action. When liquid is confined within a capillary, any contact angle other than 90° will imply either a concave (acute contact angle) or a convex (obtuse contact angle) curvature of its surface. In the concave case, the result is a pressure differential favouring ingress of liquid further along the capillary. If the capillary is oriented vertically,

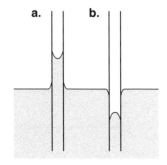

Fig. 1.9 Capillary action in cases with (a) acute contact angle, and (b) obtuse contact angle.

for example, the pressure differential can support a column of liquid against the opposing force of gravity (Fig. 1.9a). In the convex case, on the other hand, the pressure differential tends to expel liquid from the capillary, potentially preventing the passage of liquid through narrow pores even in the face of an externally applied pressure gradient (Fig. 1.9b). Clearly, such effects are most pronounced when the capillary is very narrow indeed (see Exercise 1.2), since this will maximise the surface curvature in either case.

Saturated vapour pressure

Another important consequence of surface curvature is its influence on saturated vapour pressure, which is to say the vapour pressure that must be maintained above the surface in order to prevent either net evaporation or net condensation. We can investigate this property by considering a large ensemble of identical liquid droplets of some arbitrary common radius, r, immersed in their own vapour. Let us assume that this system is in equilibrium, so that whatever value the vapour pressure happens to take must actually *be* the saturated vapour pressure corresponding to droplets of the radius we have chosen.

Now, let us imagine redistributing the constituent particles of one droplet equally amongst all the others, the number of droplets thereby falling by one while the common radius of the remaining droplets increases by some infinitesimal amount, dr. If we are to maintain equilibrium between the liquid and vapour phases, we must anticipate that it may be necessary to alter slightly the pressure in the vapour phase, by an amount dP_v, and that the pressure in the liquid phase may also vary, by an amount dP_l. Differentiating the Young–Laplace equation (Eqn. 1.18) then implies that

$$dP_l - dP_v = -2\gamma\,dr/r^2 \qquad [1.19]$$

where we assume that γ is constant.

In order to make progress, we now make use of a convenient thermodynamic relationship (one of the Maxwell relations) linking changes in chemical potential as a function of pressure to changes in volume as a function of particle quantity. Specifically, we have

$$\left(\frac{\partial\mu}{\partial P}\right)_N = \left(\frac{\partial V}{\partial N}\right)_P = \bar{V} \qquad [1.20]$$

where the symbol $\bar{V}$ indicates the molar volume of the species described, since we express N in moles. We may then write

$$d\mu = \bar{V}dP \qquad [1.21]$$

for either the liquid- or vapour-phase parts of our system.

Since we are maintaining equilibrium, it follows that the same chemical potential is common to both liquid and vapour phases, implying

$$\bar{V}_l dP_l = \bar{V}_v dP_v \qquad [1.22]$$

where $\bar{V}_l$ and $\bar{V}_v$ are the molar volumes in liquid and vapour phases respectively. Recognising that $\bar{V}_v$ is typically several orders of magnitude larger than $\bar{V}_l$, we deduce that dP_l must exceed dP_v by the same ratio, so that we may safely neglect the latter in Eqn. 1.19, yielding

$$dP_l = -2\gamma\,dr/r^2 \tag{1.23}$$

for the pressure change in the liquid phase, and hence

$$d\mu = -2\gamma\bar{V}_l dr/r^2 \tag{1.24}$$

for the change in chemical potential.

So much, then, for the infinitesimal variation of chemical potential as a function of changes in radius. Let us now use this to understand finite changes in chemical potential arising when the radius changes from a value of infinity (corresponding to droplets so large that their surfaces are flat) to some finite value (corresponding to some specific surface curvature). Integrating Eqn. 1.24, we obtain

$$\mu(a) - \mu(\infty) = \frac{2\gamma\bar{V}_l}{a} \tag{1.25}$$

where $\mu(a)$ represents the chemical potential for droplets of a particular common radius, a. Furthermore, if the vapour behaves as an ideal gas, we have

$$\mu(a) = \mu(\infty) + RT\ln[P_v(a)/P_v(\infty)] \tag{1.26}$$

with R the molar gas constant, and therefore obtain

$$\ln[P_v(a)/P_v(\infty)] = \frac{2\gamma\bar{V}_l}{RTa} \tag{1.27}$$

which is the **Kelvin equation**. The pressures appearing in this equation represent, it should be recalled, the saturated vapour pressures pertaining to droplets of a specific radius, the quantity $P_v(\infty)$ being the value usually tabulated as 'the' saturated vapour pressure in data books. Clearly, the saturated vapour pressure of small droplets may be substantially larger than the flat-surface value (Fig. 1.10), implying that they are rather more prone to evaporation than might otherwise be predicted (see Exercise 1.3).

The **Kelvin equation** describes changes in saturated vapour pressure as a function of surface curvature, modulated by temperature, surface tension, and the molar volume of the condensed phase.

Ripening and sintering

Imagine an ensemble of liquid droplets displaying a variety of radii, in contact with their own vapour at a pressure slightly higher than the standard saturated vapour pressure of the substance considered. Further, let the net rate of evaporation across all droplets equal the net rate of condensation. Are these conditions sufficient to imply that our system is in equilibrium?

In short, the answer is 'No'. The effective saturated vapour pressure for the smallest droplets will be higher than that for the largest droplets, so individual droplets will typically *not* be in equilibrium with their vapour under these

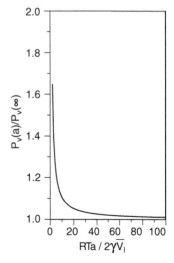

Fig. 1.10 Variation of saturated vapour pressure with droplet radius, a.

circumstances. Specifically, droplets smaller than a certain critical size will have saturated vapour pressures rather higher than the ambient vapour pressure, and hence will tend to evaporate over time, while larger droplets will tend to accumulate material via condensation from the vapour. Inevitably, therefore, the smaller droplets will gradually disappear entirely, while the average radius of the ensemble correspondingly drifts ever upwards, until the number of distinct droplets eventually dwindles down to just one (comparatively enormous) blob.

Notwithstanding our current focus upon liquid droplets, the same conclusion ultimately holds true for solid clusters too, and in either case the process of evaporation and condensation is known as **ripening**. This phenomenon is of vital interest in the context of heterogeneous catalysis, where highly reactive metals are typically prepared in nanoscopic form, either to maximise surface area or to modify electronic properties, and where an increase in average radius will radically decrease the overall catalytic activity. This, indeed, is one of the major reasons why such nanoclusters are usually immobilised on some appropriate support material that serves to impede the diffusion of particles between neighbouring nanoclusters. At the same time, the support also typically slows the rate of **sintering**—a process in which the system free energy is lowered by the migration and coalescence of entire nanoclusters, eliminating surface area abruptly with each event rather than gradually as in ripening.

> The distribution of radii in an ensemble of droplets may vary over time either by **ripening** or **sintering**. In the first case, smaller droplets evaporate while larger droplets grow through condensation. In the second, small droplets simply coalesce to form larger droplets. Both processes tend to reduce the overall surface area exposed by a given quantity of material.

1.6 Surfactants and the Gibbs isotherm

Thus far, we have developed our understanding of liquid surface thermodynamics without consideration of how the specific surface energy may vary as the chemical potentials of the system change. In our discussion of surface tension (Section 1.4) we were able to assert that chemical potentials remain constant upon perturbing either the size or shape of the surface. A number of interesting and important surface phenomena emerge, however, in consequence of chemical potentials varying with bulk concentration. One particularly profound example is the effect upon specific surface energy of **surface active agents** or (as usually contracted) **surfactants**. To fully appreciate the role these molecules play in modifying the surface, however, it will first be necessary to derive one rather fundamental isotherm.[9]

> **Surface active agents** (also known as **surfactants**) are species that can reduce the specific surface energy of a liquid.

The Gibbs isotherm

Let us recall the surface excess internal energy, defined in Eqn. 1.4, and explicitly differentiate each term to obtain

$$dU^s = TdS^s + S^s dT + \gamma dA + Ad\gamma + \sum_i (\mu_i dN_i^s + N_i^s d\mu_i) \qquad [1.28]$$

---

[9] The term 'isotherm' merely implies 'a relationship between two or more quantities that holds so long as the system temperature is constant'.

where we do not now assume the specific surface energy, γ, to be constant. We should, nevertheless, expect infinitesimal changes in the surface excess internal energy also to conform to a surface analogue of the master equation (Eqn. 1.1) so that we may write

$$dU^s = TdS^s + \gamma dA + \sum_i \mu_i \, dN_i^s \qquad [1.29]$$

in which we have inserted $dW^s = \gamma dA$ for the reversible work done in deformation of a liquid surface. Now, this last equation clearly lacks many of the terms found in the immediately preceding one, so how may we resolve this apparent contradiction?

The solution lies in the existence of an implicit relationship between thermodynamic variables, which may actually be deduced by direct comparison of Eqns. 1.28 and 1.29. It must, in fact, be the case that

$$S^s dT + A d\gamma + \sum_i N_i^s \, d\mu_i = 0 \qquad [1.30]$$

which is the surface analogue of the Gibbs–Duhem equation found in bulk thermodynamics.[10] Under conditions of constant temperature $(dT = 0)$ this simplifies (using $N_i^s = \Gamma_i A$) to the form

$$d\gamma = -\sum_i \Gamma_i \, d\mu_i \qquad [1.31]$$

otherwise known as the **Gibbs isotherm**.[11] The exact same information may also be captured if we insist that a separate equation of the form

$$\Gamma_i = -\left(\frac{\partial \gamma}{\partial \mu_i} \right) \qquad [1.32]$$

The **Gibbs isotherm** links the surface excess of different species with the change in specific surface energy as a function of their chemical potential (or bulk activity/concentration).

must hold for each and every individual species. Furthermore, another important alternative form of the Gibbs isotherm may be derived by noting the following general expression for (molar) chemical potential:

$$\mu_i = \mu_i^0 + RT \ln a_i \qquad [1.33]$$

where μ_i^0 is the chemical potential of species i in some reference state, a_i its bulk activity, and R the molar gas constant. From this, we may deduce that $d\mu_i = RTd(\ln a_i)$, and hence (after substituting into Eqn. 1.32) we conclude that

$$\Gamma_i = -\frac{1}{RT} \left(\frac{\partial \gamma}{\partial \ln a_i} \right)_{P,T} \qquad [1.34]$$

[10] For a bulk fluid, the Gibbs–Duhem equation $(SdT - VdP + \Sigma_i N_i d\mu_i = 0)$ involves P and V as the bulk pressure and volume, respectively, along with bulk entropy and particle quantities, S and N_i.

[11] The Gibbs isotherm collapses to the expected result, $d\gamma = 0$, when all chemical potentials are constant; its usefulness lies in those cases when they are not.

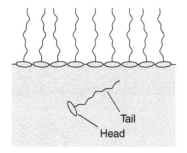

Fig. 1.11 Surfactants in excess at a liquid surface.

When the **critical micelle concentration** of a surfactant is exceeded, it will form closed structures (**micelles**) within the bulk of its host liquid, rather than increasing any further its surface excess.

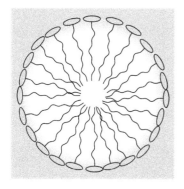

Fig. 1.12 Schematic of micelle structure.

is a valid restatement of the Gibbs isotherm. In the limit of an ideal solution, of course, activity becomes synonymous with concentration. This last expression, therefore, links the surface excess of each species to the variation of specific surface energy as a function of its bulk activity/concentration.

Salts and surfactants

The profound practical implication of the Gibbs isotherm is that a non-vanishing surface excess of any given species can modify the specific surface energy, thus affecting all the phenomena shown previously to depend upon its value (e.g. capillary action, ripening, etc.). The term 'surfactant' is usually reserved for species that accumulate at the surface (positive surface excess) and hence imply a reduction in specific surface energy as one increases the bulk concentration, but here we shall also discuss species that shun the surface (negative surface excess) tending towards an increase in specific surface energy upon increasing the bulk concentration. In general, species of each type share certain common features that allow one to predict their behaviour.

Species showing a negative surface excess, for example, are often those that possess a significant charge in solution. Within the bulk of a liquid, each ion will be surrounded by a solvation shell within which counter-ions and/ or polar molecules contrive to screen the bare charge over some characteristic length scale. Close to the surface, however, the available space for the solvation shell is curtailed on one side of the central ion, implying an energy cost amounting to a repulsive ion–surface interaction. The ions of dissolved salts will therefore typically exhibit negative surface excesses, and hence will tend to raise the specific surface energy of the solvent as their bulk concentration rises.

Species that show a positive surface excess, on the other hand, are often those that possess a distinctly elongated structure, with a 'head' end strongly attracted by the solvent and a 'tail' end strongly repelled by the same (Fig. 1.11). These surfactants may be classified into four broad categories according to the character of their head end, namely: anionic, cationic, zwitterionic, or nonionic; note that the tail end is invariably neutral. As the bulk concentration of surfactant is increased, so the positive surface excess reduces the specific surface energy of the solvent, but it should be noted that the effect cannot proceed without limit. Once the surfactant concentration exceeds a certain critical value (the **critical micelle concentration**), the formation of closed structures within the bulk solvent, known as **micelles** (Fig. 1.12), will become favourable, after which addition of further surfactant molecules will increase the number of micelles but not the excess at the free surface of the solvent. Micelles, it should be noted, can act to disperse species that might not otherwise dissolve in (or mix with) the solvent, encapsulating them inside a shell that presents highly soluble heads towards the outside world whilst the tails and insoluble species (or immiscible liquids) are discreetly sequestered within. For this reason, surfactants find practical application as soaps, detergents, and emulsifiers.

1.7 Gas/solid isotherms and relative coverage

We have seen how surfactant species accumulate in the vicinity of liquid surfaces owing to their particular electrostatic and structural properties. Key to understanding the resulting phenomena was an ability to quantify the excess amount of each species at the surface, as a function of its concentration within the bulk. Here, we shall examine how similar ideas may profitably be developed in the case of a solid surface exposed to a gaseous environment.

The Langmuir isotherm

Undoubtedly the most commonly invoked model for accumulation of material at solid surfaces during adsorption, the **Langmuir isotherm** may be derived from just four fundamental assumptions: (i) the surface is homogeneous, comprising a uniform density of identical adsorption sites; (ii) each adsorption site may be occupied by at most a single immobile adsorbate; (iii) all parameters relevant to adsorption at a given site are independent of the occupancy of neighbouring sites; and (iv) adsorption occurs directly, during initial impact of an incoming entity upon the surface, or not at all.

Invoking these assumptions, we proceed by considering the rate of adsorption onto the surface (adsorbates per site per second) given a constant flux of some particular species from the gas phase. If this species adsorbs intact, with each adsorbed entity occupying a single site, we may write

$$r_a = k_a P(1-\theta) \tag{1.35}$$

where r_a is the rate of adsorption, k_a is some appropriate rate constant for the process, P is the gas-phase pressure of the species considered, and θ represents the **relative (or fractional) coverage** of that species. That is to say, θ varies between values of zero and unity as adsorption sites gradually become filled, its intermediate values representing the fraction of sites that happen to be occupied at any given moment. The appearance of this quantity in the adsorption rate expression simply reflects the probability that an incoming entity will hit an empty site, since failure to do so will result in failure to adsorb under the Langmuir assumptions. We shall consider the nature of k_a in depth later (Section 4.2), but for now we merely note that it is generally a function of the gas temperature.

Turning next to desorption, it is natural to suppose that species having adsorbed intact from the gas phase are equally capable of returning to that phase in precisely the same form. We may therefore write

$$r_d = k_d \theta \tag{1.36}$$

for the desorption rate (adsorbates per site per second), where k_d is once again some appropriate rate constant that we shall examine in detail later (Section 4.3). For the moment, we simply remark that this rate constant is generally a function of the surface temperature. The explicit proportionality between relative coverage and desorption rate is justified simply by the quantity of adsorbates present at the surface.

The **Langmuir isotherm** predicts relative surface coverage at a solid surface as a function of the gas-phase pressure of an adsorbing species, subject to assumptions of surface homogeneity, single site-occupancy, non-interacting adsorbates, and direct adsorption.

Relative (or fractional) coverage quantifies the surface concentration of a species normalised against the maximum concentration that may be achieved on a given solid surface in a single layer.

Having separately defined both adsorption and desorption rates, it only remains to insist that these must be equal under equilibrium conditions. That is, we require

$$\frac{d\theta}{dt} = r_a - r_d = 0 \qquad [1.37]$$

with t representing time, from which we deduce

$$k_a P(1-\theta) = k_d \theta \qquad [1.38]$$

and hence

$$\theta = \frac{bP}{1+bP} \qquad [1.39]$$

where $b = k_a/k_d$ is a function of the common temperature shared by both gas and surface in this situation.[12] This, then, is the Langmuir isotherm for intact adsorption, describing the variation of relative coverage as a function of pressure at some fixed temperature (Langmuir, 1918). A selection of such isotherms for a few different values of b is depicted in Fig. 1.13a, and two limiting behaviours are clear in all cases. Firstly, in the limit of low pressure, the isotherm of Eqn. 1.39 reduces to the form $\theta = bP$ and the relative coverage is seen to vary linearly with pressure. Secondly, in the limit of high pressure, Eqn. 1.39 reduces to the form $\theta = 1$ and the relative coverage is seen to saturate with complete occupancy of all sites.

It ought to be noted that it is not always obvious, when looking at a plot of θ versus P, whether the curve obtained is truly Langmuirian in nature. Accordingly, it is common to invert Eqn. 1.39, yielding the form

$$\frac{1}{\theta} = \frac{1}{bP} + 1 \qquad [1.40]$$

so that a plot of $1/\theta$ versus $1/P$ should yield a straight line, with gradient $1/b$ and an intercept of unity. Not only does this facilitate discrimination between Langmuirian and non-Langmuirian behaviour, but it also permits the trivial extraction of a value for the constant b. Although we are deferring our discussion of k_a and k_d to a later point, it is perhaps no surprise to note here that their ratio (i.e. the constant b) is proportional to $\exp(q_a/RT)$, where q_a is the molar heat of adsorption (see Exercise 1.5).[13] Knowledge of b from a single isotherm does not, however, allow one to ascertain the precise heat of adsorption, since the constant of proportionality depends upon a temperature-independent entropic factor whose magnitude may not necessarily be known.

An extension of the Langmuir isotherm to account for dissociative adsorption is also worthy of discussion. Here, we presume that the process of adsorption

---

[12] Note that the ratio $b = k_a/k_d$ has dimensions of inverse pressure, per the definitions of the rate constants in Eqns. 1.35 and 1.36. Accordingly, the product bP is dimensionless.

[13] For adsorption occurring at constant gas-phase pressure, the heat of adsorption is given by $q_a = -\Delta H$, where ΔH (the molar adsorption enthalpy) is simply the change in enthalpy upon adsorption of one mole of gas.

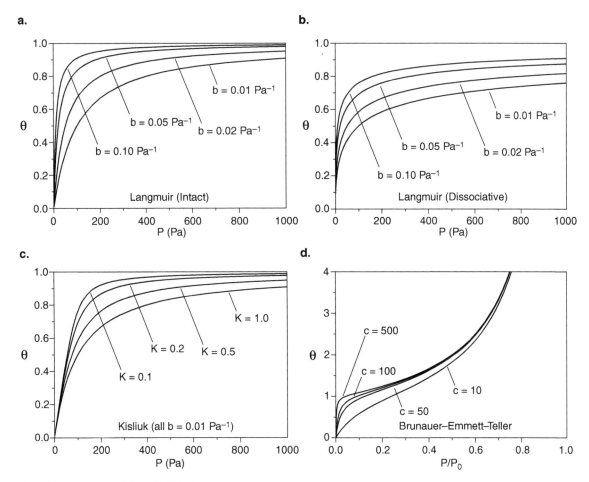

Fig. 1.13 Langmuir, Kisliuk, and BET isotherms.

requires two adjacent empty sites, to accommodate two identical fragments of the original impinging entity. Similarly, desorption must now be recombinative in nature, requiring two adjacent sites to be occupied with fragments as a prerequisite for departure from the surface. We may therefore simply write

$$k_a P(1-\theta)^2 = k_d \theta^2 \tag{1.41}$$

and hence (after some manipulation)

$$\theta = \frac{\sqrt{bP}}{1+\sqrt{bP}} \tag{1.42}$$

where $b = k_a/k_d$ is again a function of the common temperature shared by both gas and surface at equilibrium.[14] Some representative isotherms are depicted for a variety of temperatures in Fig. 1.13b.

---

[14] Clearly the rate constants now correspond to processes that are second-order in relative coverage, but their ratio retains dimensions of inverse pressure.

Finally, we state without proof that when multiple species compete for adsorption sites, the relative coverage of the ith species (within the Langmuir model) may be written as

$$\theta_i = \frac{(b_i P_i)^{n_i}}{1 + \sum_j (b_j P_j)^{n_j}}$$ [1.43]

where θ_i, b_i and P_i have the obvious interpretations, and where n_i takes the value 1 when the species in question adsorbs intact, but the value 1/2 when the relevant species adsorbs dissociatively.

The Kisliuk isotherm

One way in which it may be useful to relax the assumptions underlying the Langmuir isotherm is to permit indirect adsorption. In this model, an incoming entity need not necessarily hit an empty site immediately upon arrival at the surface, but instead can wander about for a while in a weakly bound **precursor state** that can exist even at sites where another adsorbate is already strongly bound. The lifetime of this precursor state then determines the likelihood that the entity adsorbs strongly into a suitable empty site before it can desorb. Crucial to the analysis are three different probabilities: (i) the probability that an entity visiting an empty site will strongly adsorb there, rather than desorbing or migrating elsewhere, which we shall denote p_a; (ii) the probability that an entity visiting an empty site will desorb from it, rather than migrating elsewhere or strongly adsorbing, which we shall denote p_d; and (iii) the probability that an entity visiting an already occupied site will desorb from it, rather than migrating elsewhere, which we shall denote p'_d.

> A **precursor state** is one in which an adsorbate may weakly adsorb prior to eventually either desorbing or settling into a more strongly adsorbed state.

Defining a convenient parameter

$$K = \frac{p'_d}{p_a + p_d}$$ [1.44]

it is possible to show (Kisliuk, 1957) that the rate of adsorption into empty sites may be written as

$$r_a = \frac{k_a(1-\theta)P}{1 + \theta(K-1)}$$ [1.45]

in which all other symbols retain the meanings attributed to them above. Note, though, that θ indicates the relative coverage of strongly bound adsorbates, neglecting the small fraction likely to be found in the precursor state. Equating this adsorption rate to the desorption rate used previously in deriving the Langmuir isotherm (assuming intact adsorption) we have

$$\frac{k_a(1-\theta)P}{1 + \theta(K-1)} = k_d \theta$$ [1.46]

and hence we obtain

$$(K-1)\theta^2 + (1+bP)\theta - bP = 0$$ [1.47]

as the **Kisliuk isotherm** (with $b = k_a/k_d$ as before).[15] Clearly, the condition $K = 1$ recovers the familiar Langmuir result, while smaller values permit higher relative coverages than would be achieved (at the same pressure) if only direct adsorption were allowed. Some examples are shown in Fig. 1.13c.

The **Kisliuk isotherm** predicts relative surface coverage in a similar way to the Langmuir isotherm, but relaxing the assumption of direct adsorption by granting a role to possible precursor states.

The Brunauer–Emmett–Teller isotherm

An alternative relaxation of Langmuirian assumptions is to countenance adsorption into layers on top of the one directly bonded to the surface itself. Just such a model is encapsulated in the **Brunauer–Emmett–Teller (BET) isotherm** (Brunauer et al., 1938) which we shall outline without proof.

The **Brunauer–Emmett–Teller (BET) isotherm** predicts surface coverage in a similar way to the Langmuir isotherm, but relaxing the assumption of adsorption into only a single layer by permitting the formation of adsorbed multilayers.

Underlying the model is an assumption that adsorption into all but the very first layer occurs with a common rate constant, k_a, and that desorption from all layers other than first is similarly subject to a common rate constant, k_d. Adsorption into the first layer, however, is modelled with a different rate constant, k_a', and desorption from it is similarly held to involve a distinct rate constant, k_d'. These conditions account for the fact that entities adsorbed in the first layer are bound directly to the underlying surface, while those in subsequent layers are bound only to other adsorbates. On this basis, it is possible (after tedious manipulation) to obtain

$$\theta = \frac{c(P/P_0)}{\left(1-(P/P_0)\right)\left(1+(c-1)(P/P_0)\right)} \tag{1.48}$$

where P is the gas-phase pressure, P_0 the saturated vapour pressure of the adsorbate species, and $c = k_d k_a'/k_a k_d'$ controls the relative propensity for building up coverage in the first layer compared with subsequent layers. The value of this parameter will be proportional to $\exp(q_a - q_v)/RT$, with q_a the molar heat of adsorption into the first layer, and q_v the latent heat of vapourisation of the adsorbate species. Some examples of the variation of relative coverage with pressure are shown in Fig. 1.13d, where it can be seen that for sufficiently large values of c the behaviour in the low-pressure limit approximates the Langmuir isotherm. For all values of c, however, the relative coverage eventually grows *ad infinitum* as the pressure is steadily increased.

The BET isotherm is commonly used in efforts to measure the surface area of porous materials, since it lends itself to describing the adsorption behaviour of relatively inert gases that tend not to favour special adsorption sites (e.g. argon, nitrogen, etc.). In this context, the equation of the isotherm is usually rearranged into the following form

$$\frac{P}{N(P_0-P)} = \frac{(c-1)}{cN_s}\left(\frac{P}{P_0}\right) + \frac{1}{cN_s} \tag{1.49}$$

---

[15] Evidently, it is perfectly possible to use the quoted formula to find a closed expression for θ rather than leave the isotherm as a quadratic equation, but the result is rather inelegant and confers little additional insight.

where we have introduced N_s to represent the total quantity of adsorption sites at the surface, and $N = \theta N_s$ to represent the total quantity of adsorbates. From this expression it becomes clear that a plot of $P/N(P_0 - P)$ against P/P_0 ought to yield a straight line of slope $(c-1)/cN_s$ and intercept $1/cN_s$. Measurement of N (which could be deduced, for example, from changes in the mass of a sample) as a function of P, together with knowledge of P_0, will therefore allow both the constant c and the quantity of adsorption sites N_s to be obtained. Some inkling of the surface area occupied by each adsorbed entity would then permit a reasonable estimation of the total surface area from the latter value.

1.8 Heats of adsorption and lateral interactions

The isotherms discussed previously all share one important (and questionable) assumption, which is that the parameters pertaining to adsorption at a given surface site are independent of whether neighbouring sites are occupied. For instance, we have mentioned the heat of adsorption as if it were a constant for a particular adsorbate on a particular surface, with no consideration of the possibility that it may be affected by the presence of preadsorbed entities. If we are to determine how reasonable such an assumption may be, it will first be necessary to establish a reliable method by which to measure accurately the heat of adsorption as a function of relative coverage.

Isosteric heats of adsorption

The classic method for measuring heats of adsorption is based on analysis of an isostere, which is to say a dataset obtained at constant relative coverage (cf. isotherms, obtained at constant temperature). We can then make use of the relationship

$$\ln(P_2/P_1) = -\frac{q_a}{R}\left(\frac{1}{T_2} - \frac{1}{T_1}\right)$$
[1.50]

The **Clausius–Clapeyron equation** permits analysis of isosteric data (i.e. temperature/pressure pairs obtained at constant relative coverage) to calculate heats of adsorption in reversible cases.

which is a discrete form of the **Clausius–Clapeyron equation**. This expression links two separate temperature/pressure combinations, (T_1, P_1) and (T_2, P_2), with the molar heat of adsorption, on the proviso that no net adsorption or desorption occurs in taking the system from the first macrostate to the second.[16] The practical application of this relationship therefore rests upon our ability to confirm that the equilibrium surface coverage does indeed remain unchanged as we vary the gas-phase temperature and pressure together, with the resulting temperature/pressure data pairs thereby constituting an isostere (Fig. 1.14). The gradient when plotting $\ln P$ versus $1/T$ will then simply be $-q_a/R$. Note, crucially, that we do not need to ascertain the precise value of the relative coverage—all

---

[16] Derivations of the Clausius–Clapeyron equation may be found in many textbooks on thermodynamics or physical chemistry, such as Atkins et al. (2017).

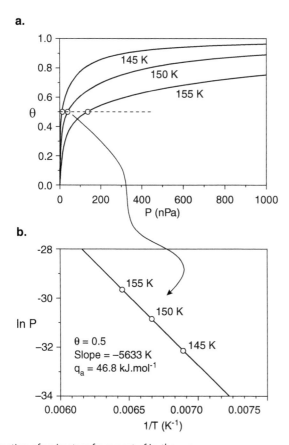

Fig. 1.14 Extraction of an isostere from a set of isotherms.

we need do is confirm, by whatever convenient means, that it does not change when we change the temperature and pressure. Having said this, if we *are* able to be specific about the coverage we use, and can moreover repeat the same trick at a number of known coverages, then we can obtain a separate heat of adsorption measurement at each one, and begin to understand its (possible) variation as a function of coverage.

Beyond the need to confirm the constancy of relative coverage during the experiment, one other drawback of the approach outlined here is the need to establish an adsorption/desorption equilibrium in the first place. Some forms of adsorption may be effectively irreversible, for example when dissociation occurs and one product desorbs, leaving an excess of the other behind on the surface, or when the surface itself is chemically altered by adsorption, as in corrosion processes. If irreversible adsorption is suspected, the isosteric approach is not appropriate, and some calorimetric method must be considered (Section 5.7). In the following discussion we shall simply assume that coverage-dependent heats of adsorption can be measured (or calculated) by some appropriate means.

Repulsive lateral interactions

In most cases, lateral interactions between adsorbates are repulsive, albeit the effect may be either relatively long-range or rather short-range. In the latter category, we note that bonding of an adsorbate to a particular adsorption site invariably modifies the local electronic structure of surface atoms immediately adjacent to that site and may partially or wholly deactivate them from participation in further adsorption. In the former category we might imagine a charged or polar adsorbate electrostatically interacting with a similarly charged or polar adsorbate situated at some considerable distance, although the surface will tend to provide dielectric screening that shortens the effective range. From the perspective of adsorption isotherms, the long- and short-range situations are significantly different.

When repulsive interactions are rather short-range in nature, their main role is to create a small exclusion zone around each adsorbate, thus dictating the effective size of adsorption site that this particular species requires. At low and moderate relative coverage, therefore, the effect on our assumption of constant adsorption heat will be negligible, since adsorbates will typically avoid each other's exclusion zones and otherwise behave as if they were adsorbed in glorious isolation from one another. Only at the very highest coverages might some deviation from the behaviours described in earlier sections occur (Fig. 1.15a).

In contrast, the existence of relatively long-range repulsive interactions can modify adsorption isotherms quite considerably, since the net heat of adsorption may be progressively lowered even when the relative coverage remains fairly moderate. In these circumstances, one may predict that adsorbates will naturally tend to maximise their separation from one another, leading to a rather uniform distribution of adsorbates, but nevertheless the effective heat of adsorption will gradually fall as the coverage increases (Fig. 1.15b).

Attractive lateral interactions

Attractive lateral interactions between adsorbates are typically only relatively short-range in nature, although they may vary in strength between rather weak and moderately strong according to their origin. In the rather weak category, we might list physical interactions such as dispersion forces, while in the moderately strong category the role of hydrogen bonds can often be of importance when suitable functional groups are present amongst the adsorbates. In either case, the effect on the heat of adsorption depends somewhat upon the surface temperature.

At sufficiently low surface temperature, when diffusion across the surface is effectively quenched, adsorbates at low relative coverage will, for the most part, remain outside the range of their attractive interactions. The effective heat of adsorption will remain largely unaltered from the value taken by an isolated adsorbate. At somewhat higher temperature, when diffusion becomes facile, adsorbates are free to coalesce into two-dimensional islands, and the heat of adsorption is augmented by the heat of interaction between the constituents

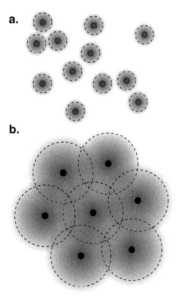

a.

b.

Fig. 1.15 Repulsive interactions affect adsorption heat only if their effective ranges overlap.

of these islands. Interestingly, however, the effective heat of adsorption remains largely independent of relative coverage in such a situation, because each newly arrived adsorbate simply adds on to the edge of an existing island (Fig. 1.16) and occupies a niche essentially identical to that of its predecessors. At still higher temperature, thermal energy amongst the adsorbates may become sufficient to preclude the formation of two-dimensional islands, and the effective heat of adsorption ought then to resemble that of an isolated adsorbate once more.

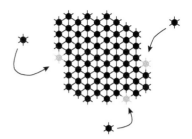

Fig. 1.16 Attractive interactions lead to island formation. Each new addition tends to find a similar local environment to the last.

1.9 Exercises

1.1 Confirm that Eqns. 1.12 and 1.13 are together equivalent to Eqn. 1.10, and that substitution of Eqn. 1.11 into Eqn. 1.10 is indeed consistent with Eqn. 1.5. Derive the condition that permits replacement of γ' with γ in Eqn. 1.13.

1.2 Water rises up a capillary of diameter 0.32 mm, achieving a height of 9.2 cm above the external level. Assuming a contact angle of zero with the capillary walls, calculate the surface tension of water. Take 997.07 kg.m^{-3} for the density of water (at 25° C) and 9.81 m.s^{-2} for the acceleration due to gravity.

1.3 In dry air (we state without proof) the instantaneous evaporation rate of a spherical water droplet is proportional both to its radius and to its saturated vapour pressure. If a single such droplet, of radius 0.01 μm, were to be mechanically dispersed into eight mutually identical smaller droplets, by what factor would the instantaneous evaporation rate of the ensemble (at 25° C) exceed that of the original droplet? Take the molar mass of water to be 18.015 g.mol^{-1}, and any other necessary parameters from Exercise 1.2.

1.4 Adsorption of N_2 on a porous graphite sample at 77 K yields the tabulated data (see margin) as a function of gas-phase pressure. Suggest which type of isotherm best fits the data, and hence estimate the sample's surface area (assuming that a single molecule occupies an area of 16Å^2). If needed, take the saturated vapour pressure of N_2 at this temperature to be 973.1 mbar.

1.5 Use the Clausius–Clapeyron equation to prove that for Langmuirian adsorption the parameter b must be proportional to $\exp(q_a/RT)$ with no other temperature dependence. Armed with this knowledge, how might one extract a value for the constant of proportionality by comparing isotherms collected at different temperatures?

Pressure (mbar)	Adsorbed Mass (mg)
9.7	13.2
19.5	19.1
29.2	22.0
48.7	24.4
68.1	26.3
97.3	28.2
146.1	31.2
194.7	33.5
243.4	35.9
292.1	39.7
389.2	46.6
486.6	57.0
729.8	112.3
778.5	136.4
827.2	186.1

1.10 Summary

- Surface free energy (surface excess grand potential) is the thermodynamic potential that a surface will minimise at equilibrium. Specific surface energy refers to the surface free energy per unit area.

- The concepts of surface tension (liquids) and surface stress (solids) both relate to work done during surface deformation. In a liquid, surface tension takes the same value as the specific surface energy and is isotropic. In a solid, surface stress may be anisotropic, and its components generally differ in value from the specific surface energy (Shuttleworth equation).

- The surface tension of a liquid may readily be measured, especially if the contact angle between the liquid and a solid surface can be quantified (Young's equation). Measuring the surface stress of a solid is less straightforward.

- Curved surfaces imply a pressure differential between concave and convex sides, related to the surface tension (Young–Laplace equation). Capillary action is one consequence, and an increase in saturated vapour pressure is another (Kelvin equation).

- Surfactants accumulate at liquid surfaces, and their surface excess is related to the specific surface energy (Gibbs isotherm). Above a critical concentration, surfactants prompt the formation of micelles.

- Adsorbates accumulating on solid surfaces may follow one of a variety of isotherms, depending upon their ability to explore the surface via precursor states, or to form multilayers (Langmuir, Kisliuk, Brunauer–Emmett–Teller isotherms).

- Heats of adsorption may be measured by isosteric means (Clausius–Clapeyron equation) and may exhibit either repulsive or attractive interactions, with consequences for adsorption isotherms.

Further reading

Atkins, P. W., De Paula, J., and Keeler, J., 2017. *Atkins' Physical Chemistry*, 11th ed. Oxford: Oxford University Press.

Brunauer, S., Emmett, P. H., and Teller, E., 1938. 'Adsorption of Gases in Multimolecular Layers'. *J. Am. Chem. Soc.* 60 (2): 309.

Gibbs, J. W., 1877. 'On the Equilibrium of Heterogeneous Substances'. *Trans. Conn. Acad.* 3: 343.

Ibach, H., 1997. 'The Role of Surface Stress in Reconstruction, Epitaxial Growth and Stabilization of Mesoscopic Structures'. *Surf. Sci. Rep.* 29 (5–6): 195.

Kisliuk, P., 1957. 'The Sticking Probabilities of Gases Chemisorbed on the Surfaces of Solids'. *J. Phys. Chem. Sol.* 3 (1–2): 95.

Langmuir, I., 1918. 'The Adsorption of Gases on Plane Surfaces of Glass, Mica and Platinum'. *J. Am. Chem. Soc.* 40 (9): 1361.

Rayleigh, Lord (Strut, J. W.), 1891. 'Surface Tension'. *Nature* 43: 437.

Shuttleworth, R., 1950. 'The Surface Tension of Solids'. *Proc. Phys. A* 63: 444.

Zisman, W. A., 1964. 'Relation of the Equilibrium Contact Angle to Liquid and Solid Constitution'. In *Advances in Chemistry: Contact Angle, Wettability, and Adhesion*, edited by F. M. Fowkes. Washington, DC: American Chemical Society.

2 Symmetry and Structure

2.1 Introduction

Having established a firm thermodynamic foundation in Chapter 1, we may now build our understanding of surface symmetry and surface structure upon a secure footing. In liquids, this amounts to little more than acknowledging the possibility of a density gradient in the surface-normal direction, accompanied by purely isotropic behaviour parallel to the plane of the surface. In solids (especially crystalline solids) the situation is far more complex. Accordingly, we shall first consider how the relative orientation of a crystalline surface determines the arrangement of its atoms or molecules, and how the underlying two-dimensional lattice must conform to one of just five symmetry types. Next, we shall introduce the seventeen space groups that encompass all possible symmetries of periodic surfaces. And finally, after introducing two widely used notations that permit easy reference to changes in surface periodicity, we will conclude by motivating the introduction of a reciprocal lattice—the key unifying concept that underlies study of all forms of surface excitation, whether electronic, vibrational, or electromagnetic in origin.

2.2 Bulk lattices and crystals

Before tackling the symmetry and structure of surfaces, we ought first to review the situation found in bulk systems.[1] Crucial to our understanding will be the observation that crystalline materials possess repetitive long-range order, and that this may best be captured mathematically through the introduction of a **real-space lattice** of points, **R**, comprising all integer linear combinations of three non-coplanar **primitive real-space lattice vectors** denoted $\mathbf{a_1}$, $\mathbf{a_2}$ and $\mathbf{a_3}$. That is, we have

$$\mathbf{R} = \alpha_1\mathbf{a_1} + \alpha_2\mathbf{a_2} + \alpha_3\mathbf{a_3} \qquad [2.1]$$

with α_i being a set of integers. Such a construct is known as a **Bravais lattice**, and may usefully be categorised according to its symmetry. In the analysis of

The **real-space lattice** of a crystalline material comprises a set of points defined by integer linear combinations of the **primitive real-space lattice vectors** (i.e. it must be a **Bravais lattice**). The lattice captures the periodic repetition of the material in three-dimensional space.

[1] A comprehensive introduction to the symmetry and structure of bulk crystals may be found in the classic text by Ashcroft and Mermin, 1976. Here, we merely summarise.

symmetry for extended objects, one must recognise not only reflection, rotation, and inversion operations (as may be found in the symmetries of molecules) but also translation, glide, and screw operations (found only in crystals). The set of all symmetry operations pertaining to an extended object is known as its **space group**, and sorted according to this attribute it may be proved that only fourteen distinct types of Bravais lattice exist. Since rather more than fourteen different space groups exist amongst the crystals found in nature, however, it is clear that the lattice alone cannot tell the whole story of crystal structure.

> The **space group** of an extended object comprises the complete set of symmetry operations that apply to it, including translation, glide and screw operations, in addition to reflection, rotation, and inversion.

Taking the lattice as merely an underlying framework, therefore, we consider the effect of placing a repeating atomic **motif** at each and every one of the lattice points, thus generating a structure that retains the long-range order inherent in a crystal, but with greater scope for variety than would be displayed by the naked lattice. In doing so, motifs consisting of only a single atom necessarily preserve all the symmetry operations found in the space group of the lattice, but less symmetric motifs may wholly or partially invalidate some of them. The space group of the crystal may therefore happen to be only a sub-group of the lattice space group. In consequence, it turns out that 230 crystallographic space groups are to be found in nature.

> A **motif** is a set of atoms that must be reproduced at each point within the real-space lattice to generate the full structure of a crystalline material. The motif can be as simple as a single atom, but may be much more complex.

As a final comment on bulk symmetry, it often proves convenient to define the so-called **point group of the space group**, which is formed by retaining all the reflection, rotation, and inversion operations found within a crystal's space group, whilst converting any glide or screw operations into (respectively) reflection and rotation operations, discarding entirely any purely translational operations. The resulting symmetry group (often described simply as the crystal's **point group**) thus contains only point symmetry operations (those that leave at least one point in the system unmoved) but is not *quite* the same thing as the set of point operations that are symmetries of the crystal structure.[2] The significance of the point group of the space group is that it determines the macroscopic morphology of the crystal; amongst bulk crystals, just 32 such point groups may be found.

> The **point group** of a crystal (more formally the **point group of the space group**) is formed by replacing glide operations in its space group with reflection operations; screw operations with rotation operations; and discarding translation operations altogether.

It is beyond our present purposes to include a comprehensive review of bulk crystal structures, but it may nevertheless be worthwhile to summarise briefly a small selection of commonly encountered ones. For example, many metals and semiconductors adopt structures based on cubic lattices, and some familiarity with at least these cases is undoubtedly desirable. Let us begin, therefore, with the simple cubic lattice, defined by the primitive real-space lattice vectors

$$\mathbf{a_1} = a\hat{\mathbf{x}} \qquad \mathbf{a_2} = a\hat{\mathbf{y}} \qquad \mathbf{a_3} = a\hat{\mathbf{z}} \qquad\qquad [2.2]$$

where $\hat{\mathbf{x}}$, $\hat{\mathbf{y}}$ and $\hat{\mathbf{z}}$ are a set of Cartesian unit vectors, and where the scaling factor a controls the spacing of lattice points. As depicted in Fig. 2.1a, it is

---

[2] Where the space group of a crystal contains glide or screw operations, the point group of the space group may contain reflection or rotation operations that do not transform the crystal structure into itself.

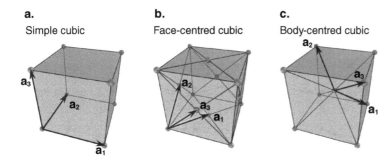

Fig. 2.1 Cubic lattices, with primitive lattice vectors marked. The cell shown in the simple cubic case is primitive; those shown in the face-centred and body-centred cubic cases are conventional.

evident that these primitive lattice vectors span a cubic region of space, which is referred to as the **primitive unit cell** of the lattice. It is worth noting that different choices for $\mathbf{a_1}$, $\mathbf{a_2}$, and $\mathbf{a_3}$ could have been made, describing precisely the same simple cubic lattice, but resulting in different (equally valid) primitive unit cells. The choice suggested in Eqn. 2.2 does, however, possess the significant merit of spanning a primitive unit cell whose point symmetry matches that of the lattice, and is therefore to be preferred under most circumstances. Attaching a single-atom motif to each lattice point results in the so-called simple cubic structure, but in fact only one element (polonium) adopts this arrangement under ambient conditions in nature. More complex crystal structures may, however, be based upon a combination of the simple cubic lattice with a multi-atom motif (e.g. caesium chloride). In any case, the primitive unit cell, together with its motif, may be tessellated to reproduce the entire crystal structure.

The full structure of a crystal may be generated by tessellating copies of the **primitive unit cell** and its contents. Any space spanned by a set of primitive real-space lattice vectors constitutes such a cell, but it is customary to choose the most symmetric amongst multiple possibilities.

If we now consider the face-centred cubic lattice, its primitive lattice vectors are usually chosen as

$$\mathbf{a_1} = (a/2)[\mathbf{\hat{x}} + \mathbf{\hat{y}}] \qquad \mathbf{a_2} = (a/2)[\mathbf{\hat{y}} + \mathbf{\hat{z}}] \qquad \mathbf{a_3} = (a/2)[\mathbf{\hat{z}} + \mathbf{\hat{x}}] \qquad [2.3]$$

and are shown in Fig. 2.1b. Note, however, that although these do indeed span the highest-symmetry primitive unit cell that may be achieved, this shape actually falls short of reflecting the full point symmetry of the lattice. Accordingly, it is often preferable to make use of a non-primitive unit cell (known as the **conventional unit cell**) spanned by the lattice vectors given in Eqn. 2.2. These are *not* primitive lattice vectors for the face-centred cubic lattice, and the conventional unit cell necessarily contains *four* lattice points, not just one. The face-centred cubic structure (obtained by attaching a single-atom motif to each lattice point) is adopted by several important metals, including aluminium, platinum, palladium, nickel, etc. Moreover, several other elements and compounds adopt crystal structures that are based upon the face-centred cubic lattice but with multi-atom motifs; for example, carbon, silicon, germanium, etc. (in the diamond structure), and zinc sulphide, gallium arsenide, indium phosphide, etc. (in the zincblende structure).

When the most symmetric primitive unit cell does not display the full symmetry of the real-space lattice, a non-primitive **conventional unit cell** displaying the full symmetry may offer a more convenient alternative.

Similarly, the body-centred cubic lattice is usually considered to have primitive lattice vectors given by

$$\mathbf{a_1} = (a/2)[\hat{\mathbf{x}}+ \hat{\mathbf{y}}- \hat{\mathbf{z}}] \qquad \mathbf{a_2} = (a/2)[\hat{\mathbf{y}}+ \hat{\mathbf{z}}- \hat{\mathbf{x}}] \qquad \mathbf{a_3} = (a/2)[\hat{\mathbf{z}}+ \hat{\mathbf{x}}- \hat{\mathbf{y}}] \qquad [2.4]$$

as shown in Fig. 2.1c. Once again, although these span the highest-symmetry primitive unit cell, it does not display the full point symmetry of the lattice. The lattice vectors given in Eqn. 2.2 are therefore often used in describing this lattice too, spanning a conventional unit cell containing *two* lattice points instead of one. The body-centred cubic structure (obtained by attaching a single-atom motif to each lattice point) is adopted by metals such as iron, molybdenum, tungsten, etc.

2.3 Miller indices and ideal surfaces

In defining a surface with reference to its underlying bulk, the relative orientation is usually expressed by means of three integers known as **Miller indices**.[3] Their use is most easily explained by example.

Consider a bulk crystal based upon a simple cubic lattice, with unit cell of side-length a. Now, given a set of Miller indices h, k, and l, we mark positions a/h, a/k, and a/l, along the three orthogonal axes spanning the unit cell, relative to an arbitrarily chosen corner. The plane that passes through these three positions is then denoted the (hkl) plane, and the corresponding (hkl) surface is realised by simply deleting all the atoms on the far side of this plane from the original corner. At this stage, we shall maintain the fiction that the remaining atoms do not move from their bulk positions, and the resulting structure is referred to as the **ideal surface** (deviations from which will be discussed in Section 2.4). Fig. 2.2 shows the concept in practice for the (213) surface, but other combinations may readily be envisaged. Note that a Miller index of zero implies that the surface plane cuts the corresponding axis at infinity, and that negative Miller indices (denoted by means of an overline) imply that the surface plane cuts on the opposite side of the original corner. Note also that it is almost universal practice to eliminate any common factors amongst the Miller indices (i.e. divide through by the greatest common denominator) since this simplifies the notation without affecting the surface orientation.

In general, Miller indices for crystals based on other lattices are defined in terms of fractional displacements along their respective primitive real-space lattice vectors, but an important exception is made when the lattice is of face-centred or body-centred cubic type. Here, the indices are invariably taken

Miller indices provide a neat way of specifying surface orientation, by means of an ordered trio of integers.

An **ideal surface** is one whose atoms are not displaced from the positions they would have occupied in the bulk material. Purely an abstraction, such a surface provides a useful benchmark against which to gauge various forms of non-ideality.

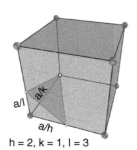

$h = 2, k = 1, l = 3$

Fig. 2.2 Miller indices relative to lattice points.

3 A four-index notation is sometimes employed for crystals of hexagonal symmetry, but we shall not discuss this here. For details, see Jenkins and Pratt, 2007.

to be defined relative to the non-primitive orthogonal lattice vectors spanning their conventional unit cells (Fig. 2.1) for essentially historical reasons.

When considering particularly high-symmetry crystals, it requires little thought to realise that several sets of Miller indices may refer to essentially identical surfaces. For example, the (213) surface of a simple cubic crystal has precisely the same structure as the (132) surface of the same material, differing only in their orientation. The $(21\bar{3})$ surface differs a little more, having a structure that turns out to be the mirror image of the other two surfaces, but is otherwise identical. To capture this succinctly, the notation {hkl} is used to denote 'the set of surfaces containing the (hkl) surface and all other surfaces interrelated by bulk point symmetry operations'. When the parent bulk crystal's point group (i.e. the point group of its space group) is cubic, the full set of surfaces implied by {hkl} may be generated simply by considering all possible permutations of order and sign for the indices. When the parent bulk crystal is rather less symmetric, however, more care must be taken. For example, the (100), (010) and (001) surfaces of a crystal structure based upon an orthorhombic lattice would not be related to one another via any combination of bulk point symmetry operations.

Surface symmetry

Having specified the orientation of our surface by means of its Miller indices, it is natural to enquire after its inherent symmetry. As a truncated version of the bulk material, it will necessarily display crystalline periodicity in the two surface-parallel dimensions, and hence may be described by means of a two-dimensional lattice of its own. That is, we have

$$\mathbf{R} = \alpha_1\mathbf{a}_1 + \alpha_2\mathbf{a}_2 \qquad [2.5]$$

where α_i represents a pair of integers, and where $\mathbf{a}_1$ and $\mathbf{a}_2$ should now be understood as the primitive real-space lattice vectors of the surface (not of the bulk). There is no longer any third primitive lattice vector, $\mathbf{a}_3$, but the outward-directed surface-normal unit vector, $\hat{\mathbf{n}}$, becomes important instead.

Now, within two dimensions, only five Bravais lattices may be distinguished on the basis of their space groups (down from fourteen in three dimensions) namely the **square**, **triangular**, **rectangular**, **rhombic**, and **oblique** lattices (shown in Fig. 2.3 with primitive cells of maximal symmetry). All crystalline surfaces must conform to one of these lattice types, with any further distinction between them stemming from different motifs attached to the lattice points. The scope for symmetry reduction implied by the motif is, however, more limited than in the bulk, leading to a total of just seventeen possible space groups (down from 230 in three dimensions).

Individual space groups may conveniently be labelled within the **Hermann-Mauguin notation**, which provides a string of up to four letters and numbers that indicate the defining symmetry operations of the group in

In two dimensions, only five types of lattice are possible: **square, triangular, rectangular, rhombic**, and **oblique**.

The **Hermann-Mauguin notation** provides a convenient descriptive approach to the labelling of different space groups.

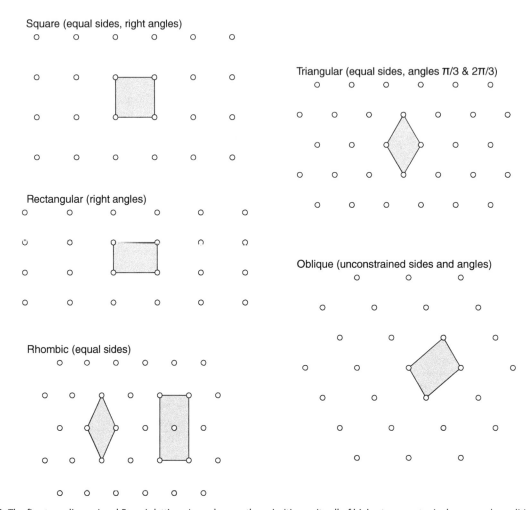

Fig. 2.3 The five two-dimensional Bravais lattices. In each case, the primitive unit cell of highest symmetry is shown, and conditions on sides/angles are given. A conventional (but non-primitive) unit cell is also shown in the rhombic case.

question.[4] The two-dimensional version of the notation begins with a lower-case letter *p* or *c*, specifying that the underlying lattice is either primitive (implying square, triangular, rectangular, oblique) or centred (implying rhombic).[5] Next follows a number, *n*, corresponding to the highest-order rotational axis 1-, 2-, 3-, 4-, or 6-fold). The final one or two positions (if utilised) then indicate the existence (or otherwise) of symmetrically distinct mirror or glide

[4] For a more detailed overview of the topics discussed here, the interested reader is referred to Ladd, 1989.

[5] The attachment of the adjective 'centred' to the rhombic lattice stems from its occasional description in terms of a non-primitive conventional unit cell that takes the form of a rectangle with lattice points not only at its corners but also at its centre (Fig. 2.3).

planes (there can be at most two) by means of the letters *m* or *g*. If both letters are present, their order relates to the orientations of these symmetry planes relative to the lattice. Thus, for example, the space group *p4gm* is based on a primitive (in this case, square) lattice and features a four-fold rotational axis, plus glide and mirror planes in distinct orientations. The space group *c1m*, on the other hand, based upon a centred (rhombic) lattice, features no rotational symmetry and only a single mirror plane. Just two unusual cases need trouble us further here, namely the *p3m1* and *p31m* space groups, where a dummy symbol (*1*) is inserted to help distinguish between two distinct mirror-plane orientations that would otherwise be notationally identical.

The seventeen two-dimensional space groups are illustrated schematically in Fig. 2.4, where relationships between them are also highlighted. These relationships become extremely important when adsorbates form ordered superstructures on the surface, since the space group of the combined geometry must either remain identical to the ideal surface's space group or else reduce to one of its sub-groups (see Exercise 2.1). Thus, for example, a surface whose ideal space group is *p2gg* could support a post-adsorption geometry with space group *p2gg*, *p2*, *p1g* or *p1*, but never *p1m*, *p2mg*, *c2mm*, etc. The same restriction will apply when we permit deviations from ideality in the structure of the surface itself (Section 2.4).

Surface structure

Turning now to the structure and stability of surfaces, it will be recalled (Section 1.3) that the key thermodynamic imperative is minimisation of surface free energy, which is to say the surface excess grand potential, $\Phi^s = \gamma A$. Moreover, for a fixed area of surface, A, this is equivalent to a requirement that the specific surface energy, γ, is in fact minimised. Indeed, one can reliably predict the relative stability of different surfaces based upon their specific surface energies, and in simple cases some quite general observations may usefully be made.

Taking face-centred cubic metals as our test case, we begin by noting that all atoms in the bulk are surrounded by twelve nearest neighbours. At the surface, however, some atoms inevitably have fewer neighbours, and the total number of 'missing' neighbours per unit area will be approximately proportional to the specific surface energy (assuming that bonding between next-nearest and further neighbours is relatively unimportant). On this basis, surfaces of {111} type turn out to have the lowest specific surface energy, followed by those of {100} type (Fig. 2.5). Indeed, these two surfaces turn out to dominate the morphology of naturally occuring face-centred cubic metal crystallites for precisely this reason.

Notwithstanding the dominance of these so-called **flat** surface types, however, even the most carefully prepared of single-crystal samples must be anticipated to expose some defective regions. For instance, a slight misorientation of the surface plane leads to a so-called **vicinal surface**, in which large **terraces** of the ideal flat structure are occasionally interrupted by atomic-scale **steps** that accommodate the angular error. Furthermore, these steps may themselves sporadically deviate from their straight path, leading to isolated **kinks** along their length. Since

A **vicinal surface** is one that is slightly misoriented relative to a **flat** surface. They will generally feature extended **terraces**, separated at irregular intervals by **steps** that may themselves feature **kinks**.

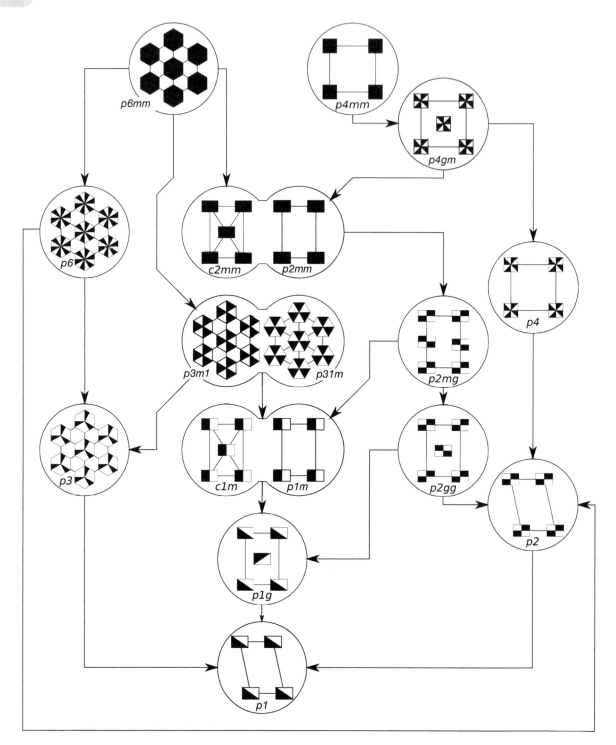

Fig. 2.4 2D Space groups, showing sub-group relationships (arrows point from group to sub-group). Note that groups *c2mm* and *p2mm* are treated as a pair, since they are mutual sub-groups of each other; groups *p3m1* and *p31m* form another such pair, as do groups *c1m* and *p1m*.

atoms at steps and kinks necessarily possess fewer neighbours than their terrace counterparts, it may well happen that they bind adsorbates more strongly and so dominate chemical reactivity (or other surface properties) despite their relative scarcity. The role of steps and kinks in surface phenomena is therefore of great importance, but the random nature of their occurence on nominally flat (or vicinal) surfaces presents an obstacle to a systematic approach towards their study.

Fortunately, the ability to artificially create **high-index surfaces**—by cutting bulk crystals at suitably obscure orientations—provides a convenient opportunity to investigate steps and kinks in a non-random setting. All such surfaces of face-centred cubic metals having precisely two Miller indices of the same magnitude exhibit a regular array of straight steps, while those whose three Miller indices are all of differing magnitudes exhibit regularly spaced steps punctuated by regularly spaced kinks. Such surfaces are of rather high energy, compared with the flat surfaces, and would not be expected to occur to any great extent in nature (see Exercise 2.2) but as a test bed for the properties of steps and kinks they are of great utility. A few examples are displayed in Fig. 2.5.

High-index surfaces are those with orientations sufficiently far from any flat surface that they feature regularly spaced steps and/or kinks.

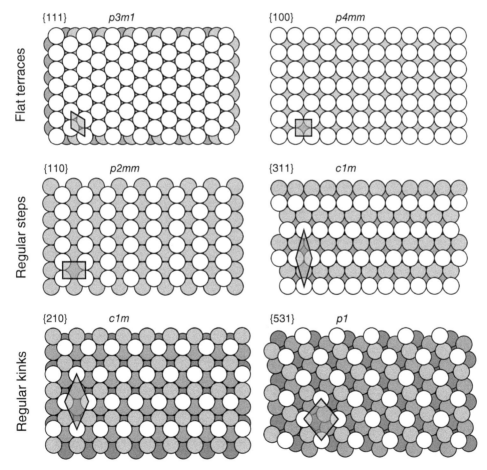

Fig. 2.5 Illustrative surfaces of the face-centred cubic structure. Primitive unit cells and space groups are indicated.

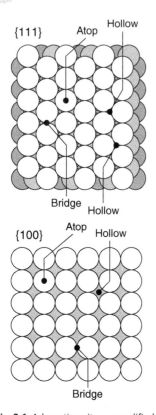

Fig. 2.6 Adsorption sites exemplified for the {111} and {100} surfaces of a face-centred cubic material.

Adsorbates can attach to the surface in a variety of different sites, of which **atop**, **bridge**, and **hollow sites** are the most frequently discussed.

When used as a unit, a **monolayer (ML)** is defined to be a coverage of one adsorbate per ideal primitive unit cell.

Surface sites

When discussing adsorption in Section 1.7, our analysis relied upon the notion that a surface would expose a set of discrete sites at which adsorption could occur, but we paid no further attention to the nature of those sites. With our new-found interest in surface structure, however, it would seem opportune to at least define some terminology with which to reference different kinds of site. On all surfaces, the possibility exists that a sufficiently small adsorbate could attach to a single atom, occupying what is generally known as an **atop site** (Fig. 2.6). Similarly, there will always be at least one alternative type that may sensibly be described as a **bridge site**, spanning two neighbouring atoms. More highly co-ordinated sites are generally described as **hollow sites**, but here we must recognise that both three-fold and four-fold variants may be found, depending upon the surface. Thus, for example, the (111) surfaces of face-centred cubic materials expose only three-fold hollow sites, while the (100) surfaces of the same materials expose only four-fold hollow sites.

For small adsorbates (single atoms, diatomic molecules, etc.) these simple sites are usually quite sufficient to describe bonding geometries, but rather larger adsorbates may imply correspondingly larger footprints on the surface. For these, it may be necessary to characterise the adsorption site as comprising several of the fundamental site types, and to speak of molecules 'centred over an atop site' or 'binding via two adjacent bridge sites' or similar. Clearly the adsorption site concept employed in developing isotherms in Chapter 1 must relate to the actual amount of surface required by the adsorbate in question, not simply to the individual atop, bridge, or hollow sites of the surface in isolation.

Surface coverage

As the foregoing discussion of adsorption sites should make clear, the maximum density of adsorbates that may be achieved on a particular surface is inversely dependent upon the area that each individual adsorbate occupies on that surface. That is, a given relative coverage (defined as per Section 1.7) may correspond to radically different adsorbate numbers depending upon the system described. A relative coverage of $\theta = 1$ for carbon monoxide on Pt{111}, for example, corresponds to 1.51×10^{19} molecules.m^{-2}, while for benzene on Ni{111} the same relative coverage amounts to 2.68×10^{18} molecules.m^{-2}; the difference arises because although the lattice constant of platinum exceeds that of nickel by 11%, the benzene molecule requires an adsorption site encompassing seven metal atoms instead of the single atom required by carbon monoxide.

For this reason, expressing the number of adsorbed molecules in terms of relative coverage may not always be the most convenient option. Instead, one often cites the number of adsorbates per ideal primitive unit cell. A coverage of precisely one adsorbate per primitive unit cell is then taken as the definition of the unit known as the **monolayer (ML)**. To continue with our prior examples, the maximum coverage for adsorbed carbon monoxide on Pt{111} would be precisely $\theta = 1$ ML, while that for benzene on Ni{111} would be $\theta = 1/7$ ML. Note that it is common (albeit highly regrettable) practice in the surface science literature to use the same symbol (θ) to represent either the dimensionless relative coverages defined in terms of adsorption sites (Section 1.7) or coverages in monolayer units defined with respect to the primitive unit cell (as here). It is important, therefore, to consider carefully the context when interpreting reported coverage values. Absolute coverages, expressed as a number of adsorbates per unit area, would be entirely unambiguous, but are only rarely cited.

2.4 Relaxation and reconstruction

Thus far, we have imagined ideal surfaces created by slicing through a bulk crystal along a plane defined by Miller indices, without allowing the atoms exposed at the surface to move in response. Plainly, however, atoms whose co-ordination number has been abruptly reduced will necessarily be subject to unbalanced forces, in this case directed predominantly so as to pull them towards the underlying bulk. The resulting movement is generally referred to as **relaxation**, and most often involves a modest, predominantly inward, shift of the outermost-layer atoms, together with progressively smaller shifts in the position of deeper layers (often with alternating sign).

In more extreme cases, relaxation may be accompanied by **reconstruction**—a term usually taken to imply that substrate atoms have moved in such a way as to reduce the symmetry of the surface from that of the ideal arrangement. The change may involve breaking point symmetries, translational symmetries, or both. In general, the space group of the surface may either remain the same as that of the non-reconstructed geometry or be reduced to one of its subgroups (as per the relationships shown in Fig. 2.4). Here, let us briefly review a few examples that illustrate some of the driving forces behind the spontaneous lowering of symmetry.

Relaxation of a surface refers to displacement of its atoms from their ideal bulk positions, without any lowering of symmetry. **Reconstruction** refers to the situation where displacement breaks the ideal surface symmetry.

Reconstruction of metal surfaces

Bonding in metals is generally rather isotropic, as evidenced by bulk structures that tend simply to maximise packing density, rather than favouring specific bond lengths or bond angles. Indeed, the face-centred cubic and hexagonal close-packed structures—both common amongst the elemental metals—represent the most efficient possible stacking of identical spheres, while the body-centred cubic structure is not far behind in this regard. Accordingly, the reconstructions of metal surfaces can often (not always) be rationalised with reference to the system's drive towards maximising packing density in the surface region.

A rather clear example of this driving force in action may be found in the so-called hex reconstruction of the Ir(100), Pt(100), and Au(100) surfaces.[6] In

---

[6] The discussion presented here is based on the reconstruction apparent on Ir(100). Strictly, the reconstructions on Pt(100) and Au(100) are very slightly different, but almost subliminally so.

the ideal (100) surface of a face-centred cubic material, atoms in the outermost layer retain eight nearest neighbours, compared with the coordination number of twelve found for atoms in all underlying layers. In the reconstructed surface, however, the outermost layer adopts a pseudo-hexagonal arrangement, in which the least coordinated atoms have only seven nearest neighbours but the most coordinated have ten. The space group of the surface changes from *p4mm* to its sub-group *p2mm*, and this justifies our identification of the associated change as a reconstruction rather than a mere relaxation. As shown in Fig. 2.7a, the outermost layer is in imperfect registry with the layer below, and the energy cost associated with this must be balanced against the gain in effective coordination number. In consequence, reconstructions of this type are not universally favourable for all face-centred cubic metals, but for the specific metals mentioned above the balance does indeed lead to pseudo-hexagonal restructuring.

An alternative situation arises on the Pt(110) and Au(110) surfaces, where the so-called missing row reconstruction is observed. In the ideal (110) surface of a face-centred cubic material, atoms in the outermost layer retain seven nearest

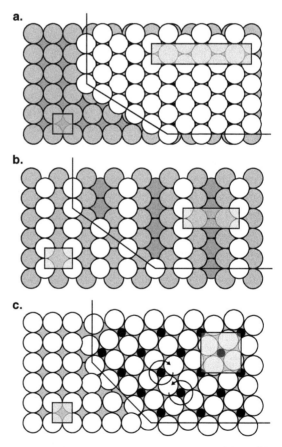

Fig. 2.7 Reconstructions on face-centred cubic metal surfaces, of (a) hex, (b) missing row, and (c) clock types. In each case, the reconstruction is shown at upper right, with the ideal surface at lower left; primitive unit cells are marked.

neighbours, while those in the second layer retain eleven, and all others possess twelve. Removing every alternate close-packed row of atoms from the outermost layer (Fig. 2.7b) means that there are only half as many atoms with just seven nearest neighbours, but the second-layer coordination number has dropped from eleven to nine, and half of the third-layer atoms now possess only eleven nearest neighbours. In fact, were one simply to assign identical weight to every nearest-neighbour interaction, the reconstructed surface would seemingly be isoenergetic with the ideal surface, so an increase in coordination numbers does not straightforwardly provide a rationale for the observed structure. In this instance, the key is that lower translational symmetry in the reconstruction permits second-layer atoms more leeway to relax laterally, in turn allowing the remaining atoms in the outermost layer to move towards the bulk. The packing density of atoms in the surface region is thus increased, in line with our expectations, even though the density of atoms in the outermost layer itself is actually reduced. Note that, in this instance, the space group of the reconstructed surface remains unchanged from that of the ideal surface (*p2mm*) but that reduction in translational symmetry (increased repeat distance in one dimension) nevertheless justifies our describing this as a reconstruction rather than a relaxation.

One final example may usefully be included, illustrating not only a case where the density of metal atoms in the outermost layer remains constant (cf. the previous two cases) but also one where the reconstruction is driven by adsorption—the so-called clock reconstruction induced by nitrogen on Ni(100). In this instance, dissociative adsorption of N_2 leads to the occupation of every other hollow site by a single nitrogen adatom, whereupon a local rotation of the neighbouring metal atoms occurs in alternately clockwise or anticlockwise fashion (Fig. 2.7c). The result is that metal atoms in the outermost layer each possess one particularly close neighbour within the same layer, while the other in-plane neighbours are slightly more distant. In this way, space is created for the nitrogen adatom to sink more deeply into the outermost layer, permitting the formation of a strong interaction with a metal atom from the second layer. The reconstruction in this instance changes the surface's space group from *p4mm* to its sub-group *p4gm*.

Reconstruction of covalent surfaces

In contrast to the metals discussed above, bonding in covalent materials may be characterised as highly directional and tending towards open structures with preferred bond lengths and bond angles. Maximisation of packing density is therefore not generally of prime importance in dictating the reconstruction of covalent surfaces. Instead, the surface seeks to minimise the number of 'dangling bonds' implied by the abrupt termination of its bulk structure. Furthermore, the surface free energy is lowered if the orbitals associated with any remaining dangling bonds are either fully occupied or entirely empty, in preference to situations where these orbitals are partially occupied.

We shall discuss the logic behind these statements rather more deeply in Section 3.3, when we consider the electronic structure of dangling bonds, so it would be premature to give specific examples here. Instead, let us simply note that the surfaces of semiconductors very frequently reconstruct by the formation of dimers, trimers and even tetramers, by which means dangling bonds may be converted to complete bonds. Surface stoichiometry may become different from that of the bulk by the accumulation or depletion of particular elements (adatoms and vacancies) and this may be beneficial in adjusting the occupancy of any remaining dangling bonds. Symmetry-lowering buckling of surface features may also be rationalised on the same basis. In comparison with metals, the range and complexity of covalent reconstructions is considerably greater.

2.5 Notation for superstructure

Wood's notation is a widely used way of recording changes in surface periodicity due to either adsorbed overlayers or reconstruction. Its simplicity means that some cases cannot be described, but for these we can make use of **matrix notation**.

As discussed above, reconstruction often leads to a reduction of translational symmetry at the surface, implying an increase in size of the primitive unit cell compared with that of the ideal geometry. Even without reconstruction, it may be that adsorbates attach to the surface in some ordered pattern whose periodic repetition is at somewhat larger scale than that of the underlying material. In either case, it will be useful to have recourse to some convenient notation that captures the change in translational symmetry. Two such notations are in common use: namely, **Wood's notation** and **matrix notation** (see Exercise 2.3). The former has the advantage of simplicity, while the latter has that of generality.

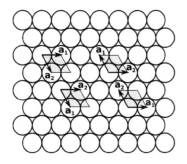

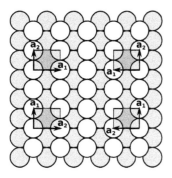

Wood's notation

The crucial prerequisite for Wood's notation (Wood, 1964) is to settle upon a reasonable choice of primitive unit cell for the ideal surface. There may be multiple competing choices, but choosing a cell shape with maximal symmetry is rarely a bad idea. Having done so, we must next decide how to define our primitive real-space lattice vectors, $\mathbf{a_1}$ and $\mathbf{a_2}$, and this is usually done so as to satisfy the following criteria:

$$|\mathbf{a_1}| \le |\mathbf{a_2}|$$
$$\mathbf{a_1}.\mathbf{a_2} \le 0$$
$$(\mathbf{a_1} \times \mathbf{a_2}).\hat{\mathbf{n}} > 0 \qquad [2.6]$$

with $\hat{\mathbf{n}}$ being the outward-directed surface-normal unit vector (Fig. 2.8). It may not always be possible to respect all three criteria, but it *is* always possible to respect at least two. When absolutely necessary to omit one, this author favours discarding the third.

Once these decisions are finalised, Wood's notation can be employed so long as a primitive unit cell of the non-ideal surface may be obtained from that of the ideal surface simply by stretching along the $\mathbf{a_1}$ and $\mathbf{a_2}$ directions (and maybe then rotating). In cases where no rotation is necessary, the notation $(m \times n)$ very simply indicates that the primitive unit cell of the non-ideal surface is obtained via stretching the ideal primitive unit cell by a factor of m in the $\mathbf{a_1}$ direction, and by a factor of n in the $\mathbf{a_2}$ direction. If a rotation must subsequently be applied, this can be incorporated in the notation $(m \times n)R\phi°$, where ϕ indicates the necessary rotational angle. A few examples are shown in Fig. 2.9. Note, in passing, that evaluating the product mn allows one to calculate instantly the area of the

Fig. 2.8 Examples of flexibility in choosing vectors spanning the primitive unit cell. Upper and lower panels respectively show {111} and {110} surfaces of a face-centred cubic material. Only the lower-right choice in each panel satisfies all three criteria of Eqn. 2.6.

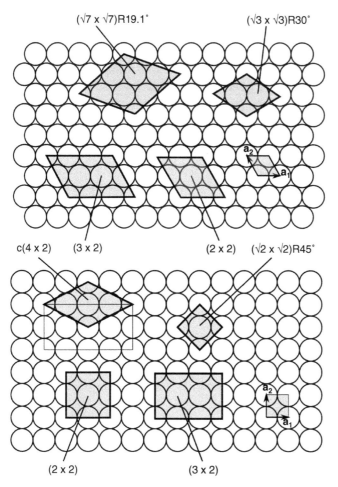

Fig. 2.9 Examples of Wood's notation, showing primitive unit cells on {111} (upper panel) and {100} (lower panel) surfaces of a face-centred cubic material.

non-ideal surface's primitive unit cell as a multiple of the ideal surface's primitive unit cell area. Thus, for example, a (2×3) unit cell would be precisely six times larger than the (1×1) unit cell belonging to the ideal surface.

On occasion, one will see the notation p$(m \times n)$ employed (with or without an appended $R\phi°$) where the 'p' prefix should be taken to represent the word 'primitive'. In this context it simply means that the unit cell of the non-ideal surface, as described by the notation, is indeed a primitive one. In the absence of any prefixed letter, however, it is usually safe to assume that the presence of 'p' is implied. In contrast, the presence of a prefixed 'c' should be taken to represent the word 'centred' and ought not to be assumed unless explicitly included in the notation. Indeed, c$(m \times n)$ implies that the unit cell obtained through scaling the ideal primitive unit cell by factors m and n is *not*, in fact, primitive. Instead, it encompasses twice the area, having lattice points located not only at its corners but also at its centre (Fig. 2.9).

Matrix notation

It is perhaps ironic that it was Wood herself who introduced (and emphasised the general superiority of) something rather akin to the modern matrix notation, in the very same article where she first set out the more limited 'shorthand' notation that now bears her name (Wood, 1964). Furthermore, the two approaches bear some similarity, in that both require one to clearly define the primitive unit cell of the ideal surface (and indeed to clarify the choice of primitive real-space lattice vectors $\mathbf{a}_1$ and $\mathbf{a}_2$ spanning it) before attempting to define the primitive unit cell of the non-ideal surface in terms of that reference. In the matrix notation, however, we simply identify the primitive real-space lattice vectors of the non-ideal surface, $\mathbf{a}'_1$ and $\mathbf{a}'_2$, as

$$\begin{pmatrix} \mathbf{a}'_1 \\ \mathbf{a}'_2 \end{pmatrix} = \begin{pmatrix} m_{11} & m_{12} \\ m_{21} & m_{22} \end{pmatrix} \begin{pmatrix} \mathbf{a}_1 \\ \mathbf{a}_2 \end{pmatrix}$$

[2.7]

albeit the matrix may be written more compactly as $(m_{11}, m_{12}; m_{21}, m_{22})$.

With suitable choices for the matrix components, it is possible for any pair of real-space lattice vectors to be indicated as spanning the primitive unit cell of the non-ideal surface. Examples are shown in Fig. 2.10, where it can be seen that cells whose shapes are indescribable within Wood's notation may be represented with ease in the matrix notation. Moreover, rotation of the non-ideal unit cell relative to the ideal unit cell is incorporated naturally into the notation, while prefixes to indicate primitive or centred cells are likewise redundant. The area of the non-ideal surface's primitive unit cell may readily be calculated (as a multiple of the ideal surface's primitive unit cell area) simply by evaluating the determinant of the matrix. Thus, for example, a (3,2; 2,3) unit cell would be precisely five times larger than the (1,0; 0,1) cell belonging to the ideal surface.

2.6 Reciprocal space

Before concluding our brief summary of surface structure and symmetry, it will prove convenient to introduce the concept of the **reciprocal lattice**. Although seemingly abstract, it does bear directly upon a number of important physical phenomena that we shall discuss in later chapters.

Bulk reciprocal lattice

The **reciprocal lattice** of a crystalline material comprises a set of points defined by integer linear combinations of the **primitive reciprocal lattice vectors**. The reciprocal lattice is useful in describing all forms of wavelike excitation within crystals.

If the real-space lattice is defined by three non-coplanar primitive lattice vectors, $\mathbf{a}_1$, $\mathbf{a}_2$, and $\mathbf{a}_3$, then one may define a corresponding set of **primitive reciprocal lattice vectors**, denoted $\mathbf{b}_1$, $\mathbf{b}_2$, and $\mathbf{b}_3$, in terms of their vector products

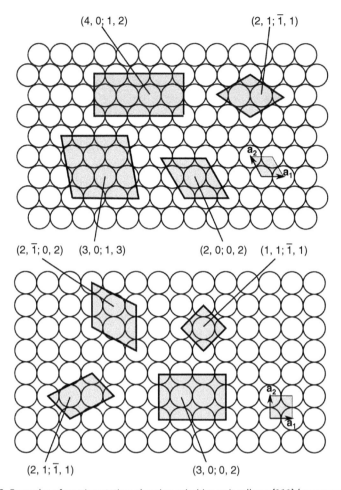

Fig. 2.10 Examples of matrix notation, showing primitive unit cells on {111} (upper panel) and {100} (lower panel) surfaces of a face-centred cubic material.

$$\mathbf{b_1} = 2\pi\frac{\mathbf{a_2} \times \mathbf{a_3}}{\mathbf{a_1}.(\mathbf{a_2} \times \mathbf{a_3})} \qquad \mathbf{b_2} = 2\pi\frac{\mathbf{a_3} \times \mathbf{a_1}}{\mathbf{a_2}.(\mathbf{a_3} \times \mathbf{a_1})} \qquad \mathbf{b_3} = 2\pi\frac{\mathbf{a_1} \times \mathbf{a_2}}{\mathbf{a_3}.(\mathbf{a_1} \times \mathbf{a_2})} \qquad [2.8]$$

and the set of all possible integer linear combinations of these constitute the reciprocal lattice. That is to say, we define a general reciprocal lattice vector, **G**, as

$$\mathbf{G} = \beta_1\mathbf{b_1} + \beta_2\mathbf{b_2} + \beta_3\mathbf{b_3} \qquad [2.9]$$

with β_i being a set of integers.

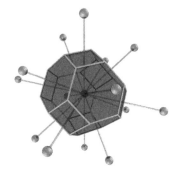

Fig. 2.11 Reciprocal lattice points for a material based upon a face-centred cubic lattice. Vectors from the origin are shown, and the 1BZ is defined by their perpendicularly bisecting planes.

The **first Brillouin zone (1BZ)** is that region of the reciprocal lattice lying closer to the origin than to any other reciprocal lattice point. Loosely speaking, it plays a similar role for the reciprocal lattice as the primitive unit cell does for the real-space lattice.

The significance of this definition of the reciprocal lattice is that it ensures the veracity of the following identity:

$$\mathbf{R.G} = 2\pi n \qquad [2.10]$$

with $\mathbf{R}$ being any real-space lattice vector, and n an integer. This, in turn, implies that any plane wave having a wavevector equal to a reciprocal lattice vector, $\mathbf{G}$, will necessarily be commensurate with the real-space lattice (in the sense that its phase will be the same at each and every real-space lattice point). The implications are of profound importance in the propagation of wavelike excitations through a crystalline system, whether those excitations be electrons, atomic vibrations, or electromagnetic radiation. In the theory of X-ray diffraction, for example, the directions in which constructive interference gives rise to intensity maxima are precisely those corresponding to the directions of reciprocal lattice vectors.

Now, just as the real-space lattice leads naturally to the concept of a unit cell that may be tessellated to fill space without overlaps or gaps, so too the reciprocal lattice gives rise to a similarly important tessellating shape, known as the **first Brillouin zone (1BZ)**. Rather than being defined as the region spanned by the reciprocal lattice vectors, however, the 1BZ is defined as the region lying closer to the origin than to any other reciprocal lattice point (Fig. 2.11). The significance of the 1BZ lies in the fact that the properties of excitations in our periodic system will simply repeat throughout reciprocal space in such a way that one need only describe them within the 1BZ to obtain a complete description of the entire system.

Surface reciprocal lattice

To define the reciprocal lattice in a two-dimensional system, we simply replace the now-redundant $\mathbf{a_3}$ lattice constant with $\hat{\mathbf{n}}$, the outward-directed surface-normal unit vector. The primitive reciprocal lattice vectors, denoted $\mathbf{b_1}$ and $\mathbf{b_2}$, then become

$$\mathbf{b_1} = 2\pi \frac{\mathbf{a_2} \times \hat{\mathbf{n}}}{|\mathbf{a_1} \times \mathbf{a_2}|} \qquad \mathbf{b_2} = 2\pi \frac{\hat{\mathbf{n}} \times \mathbf{a_1}}{|\mathbf{a_1} \times \mathbf{a_2}|} \qquad [2.11]$$

and the surface's reciprocal lattice is generated by

$$\mathbf{G} = \beta_1 \mathbf{b_1} + \beta_2 \mathbf{b_2} \qquad [2.12]$$

with the β_i once again being integers, and where $\mathbf{a_1}$ and $\mathbf{a_2}$ should be understood as the primitive real-space vectors of the surface lattice (not the bulk lattice). The identity expressed in Eqn. 2.10 holds as before, and the reciprocal lattice may be expected to once again play a key role in understanding diffraction phenomena (Section 5.2). Meanwhile, the electronic and/or vibrational properties of a surface need only be obtained within the surface 1BZ in order to completely characterise the system (Sections 3.3 and 4.4). Fig. 2.12 shows

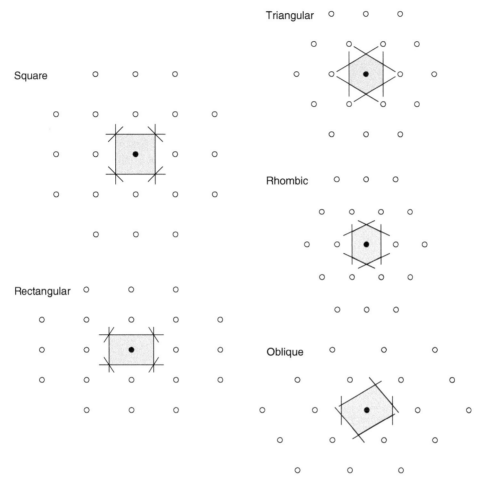

Fig. 2.12 Two-dimensional reciprocal lattices, with the 1BZ shown shaded in each case, defined by perpendicular bisectors of the shortest reciprocal lattice vectors.

the 1BZ shapes corresponding to the five basic types of surface reciprocal lattices (see Exercise 2.5).

2.7 Exercises

2.1 The space group of an ideal *fcc*-{211} surface is *p1m*. List all possible space groups that it could display after either reconstruction or adsorption. Which additional space groups would be possible for the {110} surface of the same material?

2.2 For all of the surfaces depicted in Fig. 2.5, work out the coordination number of atoms in each of the top four layers. Given that the primitive unit cell areas for {111}, {100}, {110}, {311}, {210}, and {531} surfaces are in the ratio

$3:2\sqrt{3}:2\sqrt{6}:\sqrt{33}:2\sqrt{15}:\sqrt{105}$, estimate the order of stability amongst these surfaces.

2.3 For each of the reconstructions depicted in Fig. 2.7, indicate the translational periodicity by means of both Wood's notation and matrix notation.

2.4 Potassium is adsorbed on a Ni{110} surface, forming an overlayer with c(2x2) periodicity. Assuming that no multilayer forms, suggest the most likely value of the coverage in ML units, and hence estimate the surface area per alkali metal adatom (taking the nickel lattice parameter to be 3.52 Å in its *fcc* crystal structure; the metallic and ionic radii of potassium may be taken as 2.20 Å and 1.52 Å respectively).

2.5 The (135) surface of a face-centred cubic material possesses a two-dimensional real-space lattice that may be generated from the primitive real-space lattice vectors

$$\mathbf{a_1} = \hat{\mathbf{x}} - 2\hat{\mathbf{y}} + \hat{\mathbf{z}}$$
$$\mathbf{a_2} = 2\hat{\mathbf{x}} + \hat{\mathbf{y}} - \hat{\mathbf{z}}$$

where $\hat{\mathbf{x}}$, $\hat{\mathbf{y}}$, and $\hat{\mathbf{z}}$ are unit vectors aligned with the sides of the conventional bulk unit cell. Identify the lattice type, obtain the primitive reciprocal lattice vectors, and sketch the shape of the surface 1BZ.

2.8 Summary

- The orientation of a surface relative to its parent bulk crystal may be specified by means of three integers, known as Miller indices.

- Ideal surfaces of crystalline materials are themselves crystalline, and their repetitive long-range order may be captured by means of a two-dimensional real-space lattice conforming to one of only five possible space group symmetries.

- Consideration of the motif attached to each surface lattice point enables one to categorise the symmetry of the surface structure (as opposed to that of its lattice) as conforming to one of just seventeen possible space group symmetries.

- Surface structure may be understood in terms of low-energy flat terraces separated by high-energy step and/or kink sites. Vicinal surfaces are nominally flat, but with essentially random step and/or kink distributions. High-index surfaces exhibit regular arrays of steps and/or kinks.

- A variety of fundamental adsorption sites may be exposed by a surface (e.g. atop, bridge, hollow) but large adsorbates may occupy a surface footprint comprising several such sites.

- Deviation from the ideal surface structure can involve either relaxation (in which both the space group and periodicity of the surface structure are maintained) or reconstruction (in which at least one of them is not).

- Changes in the translational periodicity of a surface—whether due to adsorption, reconstruction, or both—may be denoted by means of either Wood's notation or the more general matrix notation.

- The notion of a reciprocal lattice simplifies discussion of wavelike excitations within bulk crystalline materials, and a two-dimensional analogue of the concept may readily be defined for crystalline surfaces.

Further reading

Ashcroft, N. W., and Mermin, N. D., 1976. *Solid State Physics*. New York: Harcourt Brace Jovanovich.

Jenkins, S. J., and Pratt, S. J., 2007. 'Beyond the Surface Atlas: A Roadmap and Gazetteer for Surface Symmetry and Structure'. *Surf. Sci. Rep.* 62 (10): 373.

Ladd, M. F. C., 1989. *Symmetry in Molecules and Crystals*. Chichester: Ellis Horwood Limited.

Wood, E. A., 1964. 'Vocabulary of Surface Crystallography'. *J. Appl. Phys. A* 35 (4): 1306.

Electronic Structure

3.1 Introduction

The preceding chapters provide an overview of surface thermodynamics, symmetry, and structure—all from the perspective of the atoms and molecules involved. Conspicuously, we have avoided any discussion of the electrons that bind those atoms and molecules together, even though it is their distribution—in space and in energy—that determines the properties and processes we have expended such effort to describe. Now we shall redress the balance, switching our focus to the surface's electronic structure. Initially, our attention will light upon intrinsic properties of the surface itself, introducing a minimal model to explain such key concepts as the surface dipole and work function. Next, we shall examine the phenomenon of surface-localised electronic states, whose very existence is possible only in a two-dimensional environment, before eventually turning to the important question of how adsorbates actually bind to surfaces.

3.2 Surface dipole and work function

Let us begin by asking how electrons are distributed in the neighbourhood of the surface. Specifically, we might ask whether the proximity of the surface causes them to be distributed differently from their arrangement in the bulk material, and to answer this question it will be convenient to employ the simplest possible model for the bulk electronic structure. Accordingly, we imagine the bulk solid as comprising a hypothetical material known to theorists as **jellium**—a material in which the positive charge normally localised on discrete ion cores is smeared uniformly throughout space. In the bulk, this approximation underlies the free-electron model of electronic structure, in which the valence electrons are also uniformly distributed throughout the material. How does this situation differ at the surface?

In real materials, positive charge is localised at ion cores (comprising atomic nuclei together with the most tightly bound electrons) and the non-uniform distribution of valence electrons is largely dictated by this fact. In the imaginary material known as **jellium**, however, the positive charge of the ion cores is spread uniformly, as too are the valence electrons.

Friedel oscillations

Consider the wavefunction of a single valence electron within the bulk free-electron model. Its normalised eigenfunction takes the form of a simple plane wave

$$\psi(\mathbf{r}) = \frac{1}{\sqrt{L\Omega}}\exp(i\mathbf{k}.\mathbf{r})$$

[3.1]

where $\mathbf{r}$ is the location of the electron, $\mathbf{k}$ its wavevector, Ω the volume of a primitive unit cell, and L the number of such cells in the material. If we insist that the surface plane lies perpendicular to the $\hat{\mathbf{z}}$ direction, with vacuum above and bulk below, it will be helpful to factorise the exponential, giving

$$\psi(\mathbf{r}) = \frac{1}{\sqrt{L\Omega}}\exp(i(k_x x + k_y y))\exp(ik_z z) \qquad [3.2]$$

in which the surface-parallel (k_x, k_y) and surface-normal (k_z) components of the wavevector are separated.

Now, if the surface-normal wavevector component is positive, the wavefunction corresponds to an electron moving from the bulk towards the surface, while a negative value implies an electron moving from the surface down into the bulk. If we neglect the possibility of an electron either leaving or entering the solid through the surface, then neither a wavefunction with positive k_z nor a wavefunction with negative k_z can be adequate alone. Instead, we must construct a composite wavefunction by summing such partial waves, thus describing an electron that approaches the surface from the bulk and is then reflected back into the bulk. Conservation of momentum parallel to the surface implies that k_x and k_y must remain constant upon reflection, while conservation of energy implies that the wavevector's magnitude must remain constant.[1] Together, these conditions require that although k_z changes sign upon reflection, it must not change magnitude. Let us, therefore, set $k_z = \kappa$ (with $\kappa > 0$) for the outward-directed wavevector, and $k_z = -\kappa$ for the inward-directed wavevector, so the overall wavefunction takes the form

$$\psi(\mathbf{r}) = \frac{1}{\sqrt{L\Omega}}\exp(i(k_x x + k_y y))[\exp(i\kappa z) + \exp(-i\kappa z + i\delta)] \qquad [3.3]$$

where we have introduced an unknown phase shift, δ, between the outward- and inward-directed parts of the wavefunction, because there is no a priori reason to assume that they should be in phase.

The probability density associated with this electron can now be calculated (from the square modulus of its wavefunction) yielding

$$
\begin{aligned}
|\psi(\mathbf{r})|^2 &= \psi^*(\mathbf{r})\psi(\mathbf{r}) \\
&= \frac{2}{L\Omega}[1 + \cos(2\kappa z - \delta)]
\end{aligned}
\qquad [3.4]
$$

after some simplification. Regarding the actual value of δ, we note that the probability density associated with this electron should reduce to zero at some

---

[1] Recall that the energy of a free electron is simply proportional to the square of its wavenumber.

appropriate boundary, close to (but not necessarily coincident with) the edge of the jellium's positive charge distribution. For ease of reference, let us insist that this boundary corresponds to the height $z = 0$, whereupon we find that we must have $\delta = \pi$. We thus conclude that the valence electron density associated with a single electron takes the form

$$|\psi(\mathbf{r})|^2 = \frac{2}{L\Omega}[1 - \cos(2\kappa z)] \qquad [3.5]$$

which does indeed vanish, as required, at $z = 0$.

The total electron density at the surface is then just the sum over contributions similar to this last expression for *all* valence electrons in the system. The free-electron model dictates that these electrons occupy states having wavevectors uniformly distributed within a sphere of radius k_F (the Fermi wavenumber). Let us use this information to evaluate how many valence electrons have a particular surface-normal wavevector component, $k_z = \kappa$.

Referring to Fig. 3.1, it is evident that the number of allowed wavevectors having surface-normal components between κ and $\kappa + d\kappa$ is proportional to the area of the highlighted disk multiplied by its thickness, which is to say $\pi(k_F^2 - \kappa^2)d\kappa$. To convert this to the corresponding number of valence electrons, we need to multiply this result by the density of allowed wavevectors (which is $L\Omega/(2\pi)^3$ when employing standard boundary conditions)[2] and by the number of electrons that can share the same wavevector (which is two). The total valence electron density, $\rho_\kappa(\mathbf{r})$, contributed by these particular electrons is therefore

$$\rho_\kappa(\mathbf{r}) = 2\pi \frac{L\Omega}{(2\pi)^3}(k_F^2 - \kappa^2)|\psi(\mathbf{r})|^2 \, d\kappa$$

$$= \frac{(k_F^2 - \kappa^2)}{2\pi^2}(1 - \cos(2\kappa z))d\kappa \qquad [3.6]$$

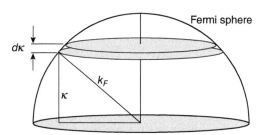

Fig. 3.1 Disc-like integration element within the Fermi sphere.

2 It is usual to apply the boundary conditions proposed by Born and Von Karman, such that wavefunctions are commensurate with the volume defined by the product $L\Omega$.

and the total valence electron density, $\rho(\mathbf{r})$, contributed by *all* of the valence electrons is thus

$$\rho(\mathbf{r}) = \int_0^{k_F} \rho_\kappa(\mathbf{r}) d\kappa$$

$$= \frac{1}{2\pi^2} \int_0^{k_F} (k_F^2 - \kappa^2)(1 - \cos(2\kappa z)) d\kappa \qquad [3.7]$$

which may be integrated and simplified (see Exercise 3.1) to obtain

$$\rho(\mathbf{r}) = \rho_0 + 3\rho_0 \left[\frac{\cos(2k_F z)}{(2k_F z)^2} - \frac{\sin(2k_F z)}{(2k_F z)^3} \right] \qquad [3.8]$$

for the valence electron density in the vicinity of the surface, where $\rho_0 = k_F^3/3\pi^2$ is the bulk valence electron density predicted by the free-electron model. The correct asymptotic behaviour in the limit $z \to -\infty$ is thus explicitly reproduced, while the vanishing of electron density in the limit $z \to 0$ is satisfied by construction.

Examining this result qualitatively, a number of features become clear (Fig. 3.2). Firstly, there is no hard division between inside and outside the solid, so far as valence electrons are concerned; the $z = 0$ plane, where the valence electron density vanishes, is well defined, but intuitively it would seem more reasonable to define the location of the 'surface' somewhat closer to the bulk (cf. the dividing plane concept, Section 1.2). Secondly, the influence of the surface on the valence electron density is felt at considerable depth, manifest as a series of periodic undulations known as **Friedel oscillations**. Their precise functional form here is dictated by the various simplifying assumptions employed in the present derivation, but similar oscillations are observed in even the most sophisticated of calculations. The common feature is that their wavenumber is twice the Fermi wavenumber, so that the characteristic wavelength of the oscillations is π/k_F. With k_F typically of the order of 10 nm^{-1} for metals, this translates to oscillations with characteristic wavelengths around 0.3 nm.

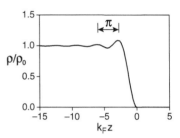

Fig. 3.2 Friedel oscillations in the valence electron density at a jellium surface.

Friedel oscillations are periodic decaying undulations in the valence electron density of a material, due to scattering from a defect of some sort. In the present case, the defect in question is the surface itself.

Surface electrostatics

We have shown that the valence electron density close to a surface does not jump discontinuously from zero to its bulk value, instead rising steeply but smoothly from zero before settling into a series of gradually diminishing oscillations. The density of positive charge, in contrast, does change abruptly within our jellium-based model, switching between its bulk value and zero on either side of a dividing plane. But where must this dividing plane be located, in relation to the plane where the valence electron density vanishes?

Let us start by evaluating the total valence charge in our system, Q^-, by integrating the valence electron density and multiplying by both the area of our surface, A, and the elementary charge, e. We thus write

$$Q^- = Ae \int_{-\infty}^0 \left[\rho_0 + 3\rho_0 \left[\frac{\cos(2k_F z)}{(2k_F z)^2} - \frac{\sin(2k_F z)}{(2k_F z)^3} \right] \right] dz \qquad [3.9]$$

and then proceed by writing an equivalent expression for the total positive jellium charge in our system, Q^+, obtained as the integral of the jellium solid's positive charge density between subtly different limits

$$Q^+ = Ae \int_{-\infty}^{z_0} \rho_0 dz$$ [3.10]

where z_0 is the location of the dividing plane.

The net charge of the surface is then given by $Q = Q^+ - Q^-$, and hence (after integrating) we obtain

$$Q = Ae\rho_0 \left(\frac{3\pi}{8k_F} + z_0 \right)$$ [3.11]

which implies

$$z_0 = -\frac{3\pi}{8k_F}$$ [3.12]

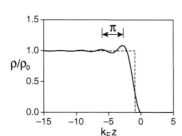

Fig. 3.3 Position of the positive jellium background charge (dashed line) relative to Friedel oscillations at a neutral surface.

as the location of the jellium edge for a neutral[3] surface ($Q = 0$). Since z_0 is plainly negative, it follows that the valence electron density extends some way beyond the dividing plane.

In Fig. 3.3, the valence electron density and positive jellium charge density are plotted together, with the appropriate spatial offset from Eqn. 3.12. Clearly there is considerable spatial separation of charge, even though the integrated net charge is zero, and this implies the existence of a non-zero dipole moment. Within the free-electron model, the dipole moment per unit area turns out to be proportional to the Fermi wavenumber, amounting to approximately $0.9\,\mathrm{D.nm}^{-2}$ when $k_F = 10\,\mathrm{nm}^{-1}$, the sign being always negative.[4]

Potential energies

Imagine a single electron, located infinitely far above the surface. In the strict absence of any other particles and of any stray electromagnetic fields, the potential energy of this electron will be essentially independent of position. We shall identify this as the **vacuum potential energy**. The total energy of the electron will likely exceed this level, because it may possess some kinetic energy, but (classically) it can never be less.

Next, imagine that this electron passes through the surface region and into the bulk. In the vicinity of the surface itself, its potential energy will vary, but in the limit of infinite depth it must once more become independent

The electrostatic properties of surfaces are highly dependant upon the **vacuum potential energy** (possessed by an electron when far outside the surface) and the **bulk potential energy** (possessed by the electron when deep inside the surface).

---

[3] In this book we shall assume our surfaces to be neutral, but note that this may not always hold true. Exceptions will include, for example, surfaces subject to an electric field, or the surfaces of insulating materials hosting adsorbed ions.

[4] The Debye, D, is a non-SI unit of dipole moment, equivalent to $3.33564 \times 10^{-30}\,\mathrm{C.m}$. For reference, the dipole moment of a single water molecule is 1.85 D.

a.   **b.**

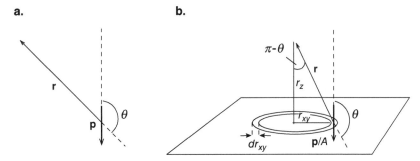

Fig. 3.4 Geometry for (a) an isolated dipole, and (b) a surface dipole sheet.

of position, attaining a value that we shall identify as the **bulk potential energy**. We expect that this should be lower than the vacuum potential energy, otherwise there would be no tendency for electrons to remain within the solid.

Now, in passing through the surface, from the vacuum potential energy to the bulk potential energy, all that has happened (from a purely electrostatic perspective) is that the electron has passed from one side of the surface dipole to the other. We know that the potential energy must change more-or-less smoothly between its bulk and vacuum values whilst traversing the surface in either direction, but the exact functional form is not obvious. Can we, nevertheless, quantify the overall magnitude of the change?

Referring to Fig. 3.4a, the electrostatic potential energy of an electron located at point $\mathbf{r}$, due to a dipole moment $\mathbf{p}$ located as shown, is given by

$$-\frac{ep\cos\theta}{4\pi\varepsilon_0 r^2} \qquad\qquad [3.13]$$

where $p = |\mathbf{p}|$, $r = |\mathbf{r}|$, θ is the angle between the dipole vector and the electron's position vector, and ε_0 is the permittivity of vacuum.[5] The expression holds true so long as the point $\mathbf{r}$ lies well away from the dipolar charge distribution.

Considering Fig. 3.4b, we deduce that the contribution to the potential energy of an electron, positioned at $\mathbf{r}$, due to the marked annulus is given by

$$dV(\mathbf{r}) = -\frac{e(p/A)\cos\theta}{4\pi\varepsilon_0 r^2}\left(2\pi r_{xy}\right)dr_{xy}$$

$$= \frac{e(p/A)r_z}{4\varepsilon_0}\cdot\frac{du}{u^{3/2}} \qquad\qquad [3.14]$$

[5] See, for example, Lorrain et al., 1988.

where the simplified form recognises that $\cos\theta = -\cos(\pi - \theta) = -r_z/r$ and makes use of the substitution $u = r^2 = r_{xy}^2 + r_z^2$ (implying $du = 2r_{xy}dr_{xy}$ with fixed r_z). Integrating over the infinite two-dimensional plane one obtains

$$V(\mathbf{r}) = \frac{e(p/A)r_z}{4\varepsilon_0} \int_{r_z^2}^{\infty} \frac{du}{u^{3/2}}$$
$$= \frac{e(p/A)}{2\varepsilon_0}$$

[3.15]

showing that the potential energy is independent of distance from the surface. By symmetry, the potential energy on the other side of the surface has the same magnitude but opposite sign. Accordingly, the change in potential energy going from the bulk side of the surface dipole to the vacuum side is just

$$\Delta V = \frac{e(p/A)}{\varepsilon_0}$$

[3.16]

which for a material with a surface dipole per unit area around $0.9\,\text{D.nm}^{-2}$ (as discussed above) would constitute a potential energy barrier of the order of 0.3 eV against the electron leaving the solid through the surface. Note, however, that the surface dipole, and hence the potential energy barrier, may be altered by relaxation, reconstruction, or the presence of adsorbates (see Section 3.4 and Exercise 3.2).

Work function

In truth, the classical electrostatic picture described above does not tell the whole story. At minimum, one should also consider the exchange contribution (arising from quantum effects associated with the fermionic nature of electrons) and the correlation contribution (arising from dynamic effects associated with motion of electrons) to the bulk potential. Including these additional effects would substantially increase the barrier that holds electrons within the bulk. Nevertheless, while the potential energy difference caused by the surface dipole may not be the only contribution to the whole, it does uniquely have the distinction of being sensitive to the precise behaviour of electrons in the vicinity of the surface.

Now, an electron lying deep within the bulk of a solid may have some kinetic energy in addition to its potential energy. Indeed, the valence electrons will collectively exhibit a range of kinetic energies spanning from zero up to some well-defined maximum value. There thus exists a Fermi level, lying some way above the bulk potential energy on a total energy diagram (see Fig. 3.5), below which all electronic states are occupied and above which none are.[6] So long as the Fermi level lies below the vacuum potential energy, the electrons will remain bound within the solid. Moreover, the energy difference between the Fermi level and the vacuum potential energy represents the minimum energy required to

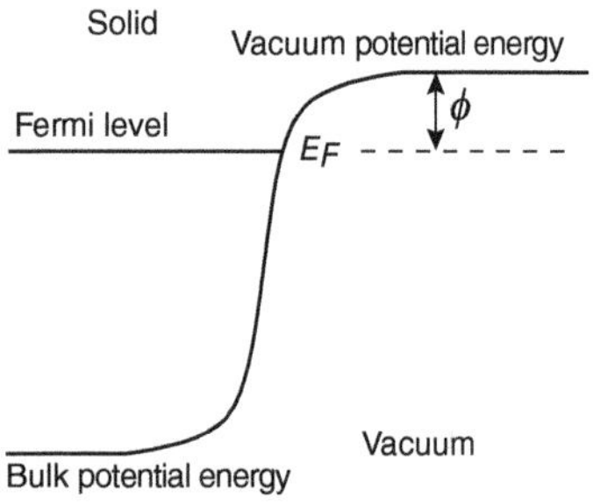

Fig. 3.5 Variation of an electron's potential energy in the vicinity of a surface.

[6] Our treatment assumes an abrupt transition from occupied to unoccupied states at the Fermi level. This is only strictly accurate at absolute zero temperature, but thermal blurring of the transition will only become important at high temperatures.

liberate an electron from the solid, and is known as the **work function** (usually given the symbol ϕ). The vast majority of metals have work functions in the range 2–6 eV, with the highly reactive alkali and alkaline-earth metals being found generally in the lower half of that range, and the less reactive noble and coinage metals being found in the upper half.

The work function is of particular importance in the phenomenon of photo-emission, whereby an electron is emitted from a material upon absorption of a single photon (the probability of simultaneous absorption of multiple photons being negligible at moderate flux). It is clear that the photon must possess an energy at least equal to the work function if it is to eject the electron, and hence that there exists a threshold frequency below which photoemission cannot oc-cur. We shall explore this more fully in Section 5.4, where we describe the utility of photoemission spectroscopy for measurement of surface electronic structure.

The **work function** of a surface is the minimum energy that must be supplied to raise an electron from the Fermi level to the vacuum potential energy, permit-ting it to be liberated from the material in a photoemission event.

Orientation-dependence of work function

As we have seen, the work function of a free-electron metal arises from a bulk contribution to the potential energy barrier (the exchange-correlation potential) added to a surface contribution (the electrostatic effect of the surface dipole) minus the Fermi energy (measured relative to the bulk potential energy). In a real material, however, the positive charge of the system is localised within ion cores, rather than smeared out as in jellium, and this can modify both the bulk and the surface contributions to the potential energy barrier, as well as the Fermi energy. Furthermore, discarding the jellium model implies acceptance that de-tails of the valence electron distribution may vary from one surface to another, not to mention that the precise locations of the ion cores most certainly will vary as well. Accordingly, we expect the surface dipole of any given material to differ from one surface facet to another, and so too must the work function.

At first glance, that last statement sounds problematic, especially when con-sidering a metal that has been cut so as to expose a variety of different crystal-line facets. We know that the Fermi level must be identical at all points within a conducting material (else the valence electrons would redistribute themselves until it was), so a difference in the work function between one facet and anoth-er would imply that the vacuum potential energy attained by a photoemitted electron must differ depending upon the facet from which it was emitted. But in our earlier definition of the work function, we started from the premise that it was possible to define a unique vacuum potential energy, so what is going on?

The answer lies in the fact that we have thus far assumed a surface of infinite lateral extent, whereas a sample featuring many facets must be a finite object. This means that the electrostatic potential energy of an electron ejected from one of its surfaces never *quite* becomes constant, or equivalently that the vacu-um region above that surface is never *entirely* devoid of electric fields. The defi-nition of the work function given above thus recognises the dominant contribu-tion of the nearest facet, but not the subtle influence of distant facets. Accurate measurements of work function are therefore made by detecting photoelec-trons at distances of little more than a few microns from the facet of interest.

Indeed, this more subtle understanding of the work function is essential if one is to avoid believing in the ultimate free lunch. If one could cause an electron to be photoemitted through one surface facet of a sample (having a relatively small work function) and it were then to be re-absorbed into the same sample through a surface facet of different type (having a rather larger work function) then the whole cycle might seem to liberate more energy than must be supplied, contradicting conservation of energy. The inevitable catch is that the electron must, between the photoemission and re-absorption steps, travel from a location just outside the first facet to a location just outside the second, and these two locations correspond to two different vacuum potential energies. In effect, the electron must work against an electric field caused by the difference in work functions, and although this field is microscopically small the distance moved through it is macroscopic. Needless to say, the work done against this field exactly accounts for the energy apparently generated from nothing by the difference between the photoemission and re-absorption steps.

3.3 Surface-localised electronic states

We have already seen how the distribution of electrons in the vicinity of the surface may differ markedly from that found in the bulk material, even when we do no more than allow for interference between bulk-like states heading towards and away from the surface (Section 3.2). But in certain systems, the effect of the surface can be more profound, leading to the existence of electronic states that can *only* arise at the surface and nowhere else. These surface-localised electronic states provide fascinating test cases for our understanding of electrons in reduced dimensions, and in some cases alter dramatically the physical and chemical properties of the entire surface. Here, we will introduce two common types of surface-localised electronic state that may occur in different circumstances, but before we learn to recognise these exceptional cases, we would do well to familiarise ourselves with the ordinary bulk-like states amongst which they may be hiding.

Projection of bulk states

Above, we adapted the free-electron model to describe surface electronic structure, starting with bulk wavefunctions in the form of simple plane waves. In moving beyond the jellium picture, the electronic eigenfunctions necessarily become more complicated, but nevertheless must (in crystalline solids) satisfy Bloch's theorem. That is, we have

$$\psi_{n\mathbf{k}}(\mathbf{r}) = \exp(i\mathbf{k}.\mathbf{r})u_{n\mathbf{k}}(\mathbf{r}) \qquad [3.17]$$

where $\psi_{n\mathbf{k}}(\mathbf{r})$ is the eigenfunction of the nth band and $u_{n\mathbf{k}}(\mathbf{r})$ is the corresponding Bloch function.[7] Unlike the eigenfunction, the Bloch function displays the same

---

7 For revision of concepts from bulk electronic structure, see Ashcroft and Mermin, 1976.

periodicity as the lattice (simply repeating from one primitive unit cell to the next) and longer-range variation is described by the plane-wave factor, exp($i\mathbf{k}.\mathbf{r}$).

Furthermore, it happens that one need only treat independently those states for which the wavevector, $\mathbf{k}$, lies within that well-defined region of reciprocal space known as the first Brillouin zone (1BZ). Any eigenfunction that happens to be Bloch-factorised with $\mathbf{k}$ lying outside of the 1BZ can be re-factorised in such a way that $\mathbf{k}$ lies inside that zone. The three-dimensional shape of the 1BZ depends upon the bulk material's reciprocal lattice, which in turn is dictated by its real-space lattice (see Section 2.6). Now, the surface of a crystalline material is essentially two-dimensional in nature, and its periodicity may therefore be described by means of a two-dimensional lattice, so it is reasonable to suppose that electronic states relevant to the surface might best be discussed in terms of the corresponding two-dimensional reciprocal lattice (also Section 2.6). Let us ask, therefore, how the three-dimensional reciprocal lattice of the bulk is related to the two-dimensional reciprocal lattice of the surface.

We start by imagining the bulk reciprocal lattice points projected in the surface-normal direction onto a flat two-dimensional plane. The resulting two-dimensional mesh of points is identical to the surface reciprocal lattice. We recognise the 1BZ of the surface, in the usual manner, as the region of the plane lying closer to the origin than to any other surface reciprocal lattice point (a region bounded by the bisectors of the shortest reciprocal lattice vectors). If we now imagine the bulk 1BZ projected isometrically onto the same plane, however, we find that its projection is rather larger than the surface 1BZ. Some points within the bulk 1BZ project straightforwardly onto points within the surface 1BZ,

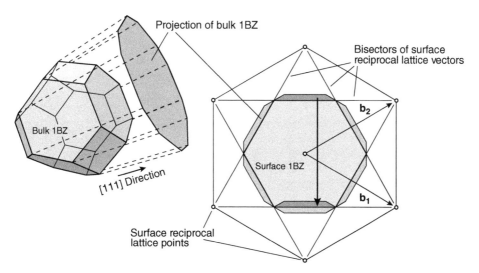

Fig. 3.6 Perspective and plan views showing the projection of the face-centred cubic 1BZ in the [111] direction (other real-space lattices and/or projection directions could equally have been chosen). One portion of the projected bulk 1BZ lying outside the surface 1BZ is shown translated through a surface reciprocal lattice vector to lie inside.

Bulk 1BZ

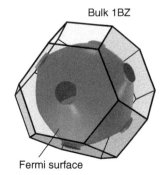

Fermi surface

Fig. 3.7 Fermi surface of Ag, shown within the bulk 1BZ.

while others project onto points that lie outside it (Fig. 3.6) but which can be shifted inside by translation through a surface reciprocal lattice vector (a symmetry operation for Bloch states).

Crucially, each point in the bulk zone maps onto a single specific point in the surface zone, but each point in the surface zone corresponds to many points within the bulk zone. If we now plot the discrete eigenvalues from each of these many bulk points against the appropriate point in the surface zone, we find that the projected bulk band structure features continuous regions of permitted and forbidden energy-wavevector space, rather than a series of discrete states.

By way of example, consider projecting the bulk band structure of Ag onto its {111} surface. In particular, note that the states near the Fermi level are essentially 5s-like, since the 4d-band is completely filled and lies at much lower energy, while the 5p-band is completely empty. These s-like valence states are rather weakly bound to the ions, and so adopt a nearly-free-electron behaviour. Consequently, the reciprocal-space surface comprising the highest occupied states (the Fermi surface) approximates a sphere, except where it approaches the edges of the bulk 1BZ and bulges outward to form a 'neck' at the centre of each hexagonal face (Fig. 3.7). When projected onto the 1BZ of a {111} surface, therefore, there exists a corresponding gap in the continuum of available states at the Fermi level close to the 1BZ centre (Fig. 3.8). Sets of states with an energy a little below the Fermi level will be distributed on a surface approximating a rather smaller sphere, with correspondingly smaller necks, and eventually, for states with low enough energy, the necks disappear altogether. In contrast, the forbidden gap around the centre of the surface 1BZ (conventionally denoted $\bar{\Gamma}$) widens at higher energies.

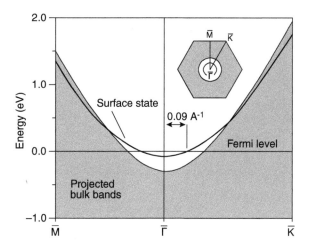

Fig. 3.8 Projected bulk band structure of Ag{111} with inset plan of the surface 1BZ. Shading indicates allowed regions for bulk states within the energy–wavevector space. The Shockley surface state is indicated with a solid line.

The example of Ag{111} thus illustrates the possibility of a forbidden gap at the Fermi level in the projected band structure of a bulk metal. Of course, the projected bulk bands must cross the Fermi level *somewhere* within the surface 1BZ, but it is not necessary for them to do so *everywhere*. Nevertheless, the presence of a forbidden gap at the Fermi level is moderately rare, occurring for Cu, Ag and Au {111} surfaces, amongst relatively few others.[8] Semiconductors and insulators, on the other hand, possess an energy gap between occupied and unoccupied states across the entire three-dimensional 1BZ in bulk, and this necessarily persists across the entire surface 1BZ.

The mere act of projecting bulk electronic states onto the surface 1BZ renders their description compatible with the constraints of a two-dimensional system, but it tells us nothing about how the behaviour of electrons lying particularly close to the surface may actually differ from those lying deep below. In reality, the different potential experienced by these near-surface electrons will feed through the Schrödinger equation and change the location of states in energy–wavevector space. Further realignment of states may occur due to changes in state occupancy related to the absence of atoms above the surface. In general, states may move either up or down in energy, and the shape of their bands may also be altered. Nevertheless, it is important to recognise that states displaced only a little from their bulk energies, and hence remaining within the allowed regions of the projected bulk band structure, will couple strongly to electronic states far below the surface. At most, they may be regarded as **surface resonances**, being only modestly localised in the surface region and possessing a non-vanishing sub-surface tail. In contrast, states that have been shifted out of the allowed regions of the projected bulk band structure altogether are incapable of coupling with bulk states, and are therefore extremely sharply localised within the top few surface layers. True **surface states** are inherently quasi-two-dimensional in nature, and constitute some of the most intriguing of all surface electronic phenomena. Moreover, when surface states are present, they often dominate the combined contribution of the bulk-like states in dictating the physical and chemical properties of the surface as a whole.

Electronic bands that lie outside of the bulk projected continuum must correspond to **surface states** that are strongly localised at the outermost layers of the material. Where such bands cross into the projected bulk states, incompletely localised **surface resonances** may occur.

Shockley states

Surface electronic states commonly arise in one of two forms (although there exist various more esoteric types that we shall ignore). The first of these—**Shockley states**—are essentially nearly-free-electron-like states that have been shifted outside the bulk-allowed regions through the effect of the surface potential.[9] A classic example is provided by Ag{111} where a parabolic band is shifted into the Fermi-level gap around the zone centre (Fig. 3.8). This parabolic band is a diagnostic feature of Shockley states, reflecting the fact that the relevant electrons

Shockley states are nearly-free-electron-like states, characterised by parabolic dispersion.

---

[8] Forbidden gaps that do not cross the Fermi level are comparatively commonplace.

[9] The author feels compelled to note that William Shockley, after whom these states are named, became notorious in his later career for promoting deeply problematic views on race and eugenics. There is a broader conversation to be had about how scientific terminology memorialises individuals whose ethics or morality may later turn out to be questionable.

behave in a nearly-free manner within the surface plane. In fact, it is possible to parametrise the dispersion with reference to an effective mass, m^*, for the surface-state electrons, obtaining

$$\varepsilon = \varepsilon_{min} + \frac{\hbar^2 k_{xy}^2}{2m^*}$$

[3.18]

where k_{xy} is the wavenumber within the xy plane of the surface, and ε_{min} is the energy at the bottom of the parabolic band (see Exercise 3.3). Although the electron density associated with this state is rather featureless within the surface plane, it is far from featureless in the surface-normal direction, being very strongly surface-localised. Notably, we can predict from the location of its band that this state will behave like a true surface state only very close to $\bar{\Gamma}$, while closer to the zone edge it will enter the allowed region of the projected bulk band structure and become a surface resonance.

The wavenumber at which the parabolic band crosses the Fermi level may be regarded as its Fermi wavenumber, k_F, and measured via Angle-Resolved Ultraviolet Photoemission Spectroscopy (Section 5.4) or through Scanning Tunnelling Microscopy (Section 5.3). In the latter case, one observes oscillations in the surface electron density due to scattering from defects (steps, adatoms, etc.) that are a close two-dimensional analogoue to the Friedel oscillations described in three dimensions above (Section 3.2) and which once again may be shown to display a characteristic wavelength of π/k_F.[10] Deliberate construction of 'quantum corrals' has even enabled the spectacular demonstration of quantum mechanical particle-in-a-box behaviour when these quasi-two-dimensional free-electron states are artificially confined (Fig. 3.9). Not only do such experiments dramatically confirm some of the most fundamental predictions of quantum mechanics, but they also provide a testbed for the investigation of open questions in many-body physics (Fiete and Heller, 2003).

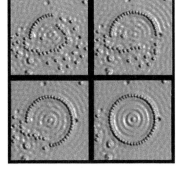

Fig. 3.9 Stepwise construction of a quantum corral from 48 Fe adatoms on a Cu{111} surface, leading to electronic standing waves (Crommie et al., 1995).

Tamm states are dangling bond states, often characterised by rather flat dispersion curves.

Tamm states

The second kind of surface state owes as much to changes in occupancy at the surface as to changes in its potential. These are the so-called 'dangling bond' states—**Tamm states**—in which localised covalent orbitals (such as one might expect to find in an insulator) become partially occupied because of the absence of atoms above the surface. In the simplest cases, bonding orbitals in homopolar covalent materials (e.g. Si, Ge, etc.) may be thought of as containing one electron from each of two neighbouring atoms. When we form a surface, the bonds left dangling upon doing so will naturally contain only a single electron if we disallow redistribution of electrons between atoms.

---

[10] It is very important to recognise that the Fermi wavenumber relevant to Friedel oscillations within the plane of the surface is that of the Shockley state whose scattering gives rise to the phenomenon, not that of the underlying bulk material. The latter is only relevant to Friedel oscillations perpendicular to the plane of the surface.

Consequently, such Tamm states must cross the Fermi level, which lies within the band gap of the bulk material.

As noted in Section 2.4, however, the surfaces of covalent materials often reconstruct, and the Tamm states provide a rationale for why this should happen. Essentially, any modification of the surface that can drive Tamm states away from the Fermi level should lower the energy—states moving up in energy necessarily become empty, and hence no longer contribute to the total electronic energy, while states moving down in energy become fully occupied and lower the potential energy as they do so. Three broad possibilities exist:

i   Neighbouring surface atoms may move towards one another, enabling dangling bonds to overlap, forming filled bonding and empty antibonding linear combinations.

ii Neighbouring surface atoms may exchange electrons, so that the dangling bond on one becomes filled, while that on another becomes empty.

iii An adsorbate may connect to the surface via one of the dangling bonds, creating filled bonding and empty antibonding linear combinations.

The first scenario is generally accompanied by a lowering of the system's translational symmetry, as symmetrically equivalent surface atoms are displaced in a manner that typically increases the repeat distance of the lattice in at least one direction; this is an example of the phenomenon known generally as a **Peierls distortion**. The second scenario need not involve any change in translational symmetry but often lowers the point symmetry, in which case it may be considered an example of the phenomenon known generally as a **Jahn–Teller distortion**. Finally, the third scenario may or may not modify the surface symmetry and is generally described under the title of **passivation**.

To illustrate all three scenarios, let us consider the classic example of the Si{001} surface, shown in a schematic side view of its unreconstructed condition in Fig. 3.10a. As depicted, each atom in the surface layer makes two complete sp^3-like bonds downwards towards the second layer (one of which obscures the other in the figure) and is conventionally regarded as possessing two sp^3-like dangling bonds pointing into empty space. The two complete bonds are filled with two electrons each, while the dangling bonds are semi-occupied, so the 1BZ of this surface features two Tamm states crossing the Fermi level. Astute readers will notice, however, that these Tamm states do not simply form a degenerate pair, as might have been expected based upon the sp^3-like picture. In fact, one of the Tamm states rehybridises with the two complete bonds, taking on sp^2-like character and dangling vertically out of the surface, consequently exhibiting very little variation in energy across the 1BZ. The remaining Tamm state then takes the form of a simple p orbital, oriented perpendicular to the plane defined by the two complete bonds, exhibiting strong variation in energy along certain directions within the 1BZ. The conventional sp^3-like dangling bonds depicted in the schematic should therefore be interpreted whilst keeping this more subtle picture firmly in mind.

Surface reconstructions may be driven by the removal of Tamm states from the vicinity of the Fermi level. Where this involves a lowering of translational symmetry we may speak of a **Peierls distortion**, and where lowering of point symmetry is involved we might describe this as a **Jahn–Teller distortion**. Tamm states may also be removed by **passivation**, involving the saturation of dangling bonds upon adsorption.

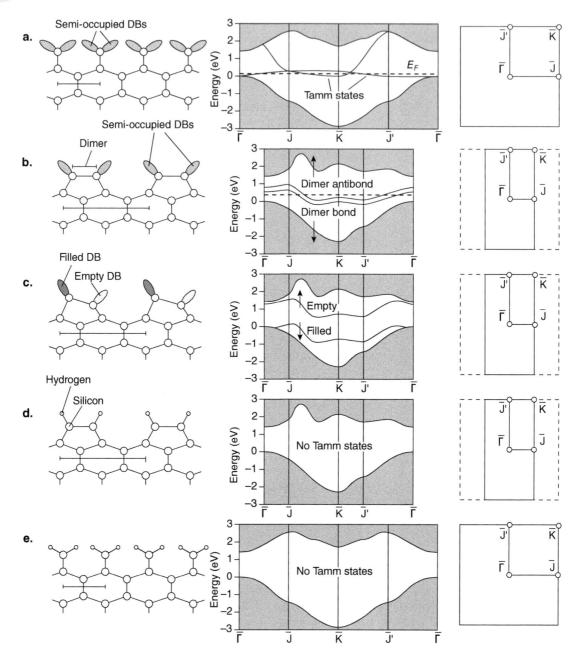

Fig. 3.10 Schematic side views of Si{001} showing (a) the ideal surface, (b) the symmetric dimer reconstruction, (c) the asymmetric dimer reconstruction, (d) the monohydride surface, and (e) the dihydride surface. Corresponding projected bulk band structures and surface states are shown alongside, with plans of the surface 1BZ.

In Fig. 3.10b, we see the result of a Peierls distortion in which neighbouring surface atoms have moved together to form dimers. The translational symmetry of the surface has been reduced, and the primitive surface unit cell has doubled in size along the direction parallel to the dimers. Accordingly, the 1BZ for the surface will have halved in size along that same direction, and states falling out-side of this smaller region must be brought inside by translation along one of the new primitive reciprocal lattice vectors. We must, therefore, account for as many as four localised states within the 1BZ of the Peierls-distorted surface, two of which correspond to linear combinations of the semi-occupied sp^3-like or-bitals depicted in the schematic (resulting in two near-degenerate Tamm states crossing the Fermi level in the band structure plot). The other two, however, now form bonding and antibonding combinations associated with dimerisation, splitting either side of the Fermi level (becoming lost amongst the projected bulk states) and lowering the overall energy.

In Fig. 3.10c, we see the result of a subsequent Jahn–Teller distortion. Here, an electron hops from one dimer atom to the other, so that half of the dangling bonds become fully occupied and the other half become entirely empty; the correspond-ing Tamm states are split either side of the Fermi level, lowering the overall energy further. The loss of symmetry in the electronic structure implies a concomitant loss of symmetry in the surface geometry, with the electron-rich dimer atom rising away from the surface and its electron-poor partner moving towards it. Some degree of rehybridisation away from sp^3-like orbitals will clearly be involved. Note, however, that for both Peierls and Jahn–Teller distortions, the electronic driving force towards symmetry breaking is offset somewhat by a build-up of energy due to straining the other bonds in the system. Accordingly, the degree of distortion in any given case will be limited by the stiffness of the crystalline structure.

If hydrogen is adsorbed on the Si{001} surface, the result depends upon the coverage achieved. At relatively low coverage (Fig. 3.10d) the preferred adsorption geometry sees the silicon dimers maintained, but the remaining sp^3-like dangling bonds are each passivated by the attachment of a single hydrogen atom. Since this leaves the surface with no partially occupied bonds, there is no longer any driv-ing force towards buckling of the dimers, which consequently retain a symmetric geometry. At higher relative coverages (Fig. 3.10e) the dimers themselves may be broken apart, with every dangling bond of the clean surface passivated by the at-tachment of a single hydrogen atom. Once again, this results in a surface devoid of any partially occupied bonds, so the driving force towards dimerisation is removed.

In general, the logic mapped out above is capable of rationalising most re-constructions observed at semiconducting surfaces.[11] One begins by identify-ing the dangling bonds found at the unreconstructed surface, then proceeds to count electrons in order to ascertain the nominal dangling bond occupancies. Finally, one considers whether passivation or symmetry-breaking distortion can achieve a situation where all Tamm states are split away from the Fermi level. In compound materials, however, the electron-counting stage is complicated by the fact that each bond in the bulk contains differing numbers of electrons from different species (e.g. in III-V semiconductors, the cation contributes $0.75e$ to each bulk bond, while the anion contributes $1.25e$). Also, the stoichiometry of surface layers may vary in pursuit of the minimum number of partially occupied dangling bonds (see Exercise 3.4).

---

[11] For a review encompassing a plethora of examples, see Srivastava, 1997.

3.4 Adsorbate-surface bonding

Atoms and molecules can stick onto solid surfaces either by formation of a strong chemical bond (along the spectrum between ionic and covalent extremes) or in the absence of a strong chemical bond (due to physical attraction involving polarisation). The former mode is known as **chemisorption** (chemical adsorption) and the latter as **physisorption** (physical adsorption). Here, we shall explore the nature of the adsorbate–surface interactions in each case.

Adsorption may occur either through the formation of a strong chemical bond (**chemisorption**) or through weak physical interactions (**physisorption**).

   In general, we shall assert that: (i) the formation of an ionic bond between two entities requires wholesale transfer of electrons from an orbital located wholly or primarily on one entity into an orbital located wholly or primarily on the other; and (ii) that the formation of a covalent bond requires transfer of electrons donated by at least one of the entities into an orbital created as a linear combination of orbitals from both entities. Each of these scenarios implies a key role for the frontier orbitals (i.e. the highest occupied and lowest unoccupied orbitals) as these are most apt either to donate or to accept electrons, or indeed to contribute to linear combinations that may be formed between adsorbate orbitals and substrate orbitals. First, however, let us consider how bonding may occur in the absence of any such chemical effects.

Physisorption

It ought to be stressed that the phenomena responsible for physisorption are ever-present, even when chemical binding (ionic or covalent) is also evident. When an adsorbate is also chemisorbed, however, most authors tend to downplay the purely physical contributions to its overall adsorption heat. This is unfortunate but entirely understandable, given that the chemical effects in these cases typically outweigh the physical by perhaps an order of magnitude. Undoubtedly, physical interactions between adsorbate and surface are most clearly seen in the case of pure physisorption, where neither ionic nor covalent bonding is able to mask their effects.[12] Broadly speaking, these physical interactions are all consequent upon the nature of any dipole that may be associated with the adsorbate.

   We have already seen (Section 3.2) that the surface naturally possesses a permanent dipole moment, responsible (alongside a significant contribution from the bulk exchange-correlation interaction) for the potential energy slope experienced by an electron as it moves through the near-surface region (Fig. 3.5). Such a slope is, of course, synonymous with the existence of an electric field in the vicinity of the surface, so let us consider how this field may interact with adsorbate dipoles.

---

[12] Systems exhibiting pure physisorption may be expected when either the adsorbate or the surface (or both) are very unreactive; for example, when the adsorbate is a noble gas atom, or when the surface is a coinage metal such as gold.

Consider first the case where an adsorbate already possesses a permanent dipole moment in its gas-phase state, and assume for simplicity that this does not change when it approaches the surface (see Exercise 3.2). When a permanent dipole enters a region of non-vanishing electric field, two things naturally occur: firstly, there is an energetic preference for the dipole to orient parallel to the field (in opposition to the surface dipole); and secondly, the dipole will be drawn towards the point of maximum field strength (the position where the potential slope is maximised). There will thus be a long-range attractive interaction between adsorbate and surface, albeit counteracted at short range by Pauli repulsion once occupied adsorbate orbitals overlap with occupied surface orbitals.

If we now consider the possibility that proximity to the surface may influence the electronic structure of the adsorbate, we should expect that the presence of an electric field will induce a dipole moment on the adsorbate, in the same sense as the preferred orientation adopted by a permanent dipole in the same field. That is, an adsorbate with a permanent gas-phase dipole will see that dipole enhanced by an induced contribution, while an adsorbate with no permanent dipole of its own will gain an induced dipole in opposition to the surface dipole. In either case, the result is a further long-range attractive interaction between the adsorbate and the surface.

Finally, we should also note the contribution of the London dispersion force between surface and adsorbate, which may be thought of as an interaction between co-induced virtual dipoles located on the two parts of our system. This long-range attraction is present regardless of whether the adsorbate possesses a static dipole moment (permanent or induced) and may constitute a significant contribution to the overall physisorption interaction in any case.

Chemisorption (ionic)

Let us turn to the case where unidirectional transfer of charge between substrate and adsorbate occurs. We will look in detail at alkali metal adsorption, where this mode of bonding is seen most clearly, but could equally have taken halogen atoms as our example (albeit with the direction of charge transfer and any consequential changes reversed).[13]

The ionisation energy of alkali metal atoms lies in the range 3.9–5.4 eV, implying that their highest (partially) occupied electronic energy level lies just that amount below the local vacuum potential energy. Since the work functions of many metals are rather larger than this, it follows that we may often be dealing with a situation where this state lies some way above the Fermi level of the surface. In Fig. 3.11, we schematically depict a typical case, indicating that the energy of the system may readily be lowered by wholesale transfer of the adsorbate's single valence electron into the continuum of unoccupied states just above the Fermi level. Two subtleties are worth noting, however, namely: (i) that

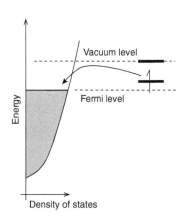

Fig. 3.11 Density of states for a solid surface (depicted as a free-electron metal for simplicity) juxtaposed with the orbital energy of the semi-occupied state of an alkali metal atom.

---

[13] Alkaline earth metals and chalcogens may also be thought of as bonding to metal surfaces in predominantly ionic fashion, although some degree of metallic bonding (in the former case) and covalent bonding (in the latter) complicates matters.

charge transfer will be kinetically hindered, by the need for electronic tunnelling across a vacuum gap, until the adsorbate approaches the surface very closely; and (ii) that as the local vacuum level experienced by the adsorbate slopes down in the vicinity of the surface, so the highest occupied adsorbate state will drop in energy and may approach the Fermi level, leading to broadening and possible incomplete charge transfer. Nevertheless, we should expect the bonding of alkali metals at surfaces to be predominantly ionic in nature (Gurney, 1935).

One particularly clear signature of ionic adsorption ought to be revealed in the work function of the surface, which we have seen is strongly related to the surface dipole. The transfer of an electron from adsorbate to surface should reduce the surface dipole moment, and hence correspondingly reduce the observed work function. We can even estimate the likely size of this effect by assuming that a single electron is transferred from each adsorbate, and that the vertical separation of the positive and negative charges thus produced is going to be roughly equal to the ionic diameter. Depending upon the size of the alkali metal ion, the charge separation is likely to be roughly 2–3 Å, so the additional dipole moment will be around $3-5 \times 10^{-29}$ C.m(10–15D) per adsorbate atom. A close-packed layer of such atoms might achieve a density close to $10^{18} \mathrm{m}^{-2}$, making a total contribution of $3-5 \times 10^{-11} \mathrm{C.m}^{-1}$(10–15 D.nm^{-2}). This would imply (using Eqn. 3.16) a drop in work function of around 3–5 eV for a complete layer.

To test this expectation, one would ideally wish to measure the work function as an alkali metal overlayer is either gradually deposited upon or removed from the surface. Results of just such an experiment are shown in Fig. 3.12, revealing a negative change in the work function as expected. Furthermore, the drop is initially fairly linear as a function of coverage, consistent with the notion that each adatom contributes a fixed dipole moment to the surface. Notably, however, the slope of the work function change gradually flattens out at a relative coverage just over $\theta = 0.5$, before the work function actually starts to rise. The generally accepted explanation for this observation (common in alkali metal adsorption) is that as coverage increases the dipoles created by individual adatoms get much closer together, destabilising them and tending to reduce the degree of charge transfer per atom. Eventually, as the adatoms start to form a close-packed layer, the adsorbate–substrate bonding takes on a more metallic character, and the work function starts to move towards a value consistent with that of the pure alkali metal surface. Nonetheless, at low coverages the alkali metal serves to significantly reduce the work function—an effect historically exploited in the design of filaments for thermionic emission of electrons (e.g. caesiated tungsten, etc.).

Chemisorption (covalent)

As described, ionic chemisorption occurs when an adsorbate's highest occupied orbital lies above the substrate's Fermi level, or equally when the adsorbate's lowest unoccupied orbital lies below the substrate's Fermi level. When the frontier orbitals of the adsorbate lie either side of the surface Fermi level, however, there is no driving force towards the wholesale transfer of electrons. In such

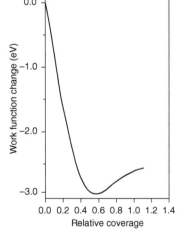

Fig. 3.12 Work function variation during adsorption of Cs on Ag{111} (after Argile and Rhead, 1988).

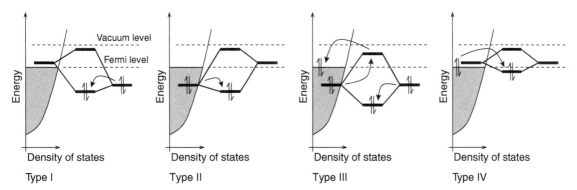

Fig. 3.13 Covalent adsorbate-substrate interactions of the extended Blyholder model. The surface electronic structure is schematised to comprise a small number of energy-localised states against a background continuum of free-electron character.

cases, chemisorption is associated with the formation of covalent linear combinations that mix orbitals of adsorbate and surface origin.

Consider the simplified energy level diagram in Fig. 3.13, which illustrates four distinct types of interaction, all of which may be relevant to covalent adsorption in greater or lesser degree. Interactions such as these were first discussed by Blyholder (1964) in the context of CO adsorption on metal surfaces, where the key molecular orbitals are 5σ (HOMO) and $2\pi^*$ (LUMO). Here, we take our cue from a slightly more nuanced approach, elucidated first by Hoffmann (1988).

Interactions of Types I and II are essentially those originally described in the Blyholder picture. In Type I (II) an occupied (unoccupied) molecular orbital forms bonding and antibonding combinations with an unoccupied (occupied) surface orbital. Of necessity, the bonding combination lies below the Fermi level, while the antibonding combination lies above. Splitting is maximised when the adsorbate orbital lies close to the surface Fermi level. The interaction involves a total of two electrons, which originally occupied either one of the adsorbate or one of the surface orbitals, and which now occupy orbitals of mixed adsorbate-substrate character. The net flow of electrons is from the molecule towards the surface for interactions of Type I, and from the surface towards the molecule for interactions of Type II. The Type I interaction is usually described as **donation** (from the molecule to the surface) and the Type II interaction as **back-donation** (from the surface back to the molecule) although in neither case is the transfer complete (since the mixed orbital involved is shared between molecule and surface).

Type III interactions arise when linear combinations are taken between an initially occupied adsorbate orbital and an initially occupied substrate orbital. If both of the resulting mixed orbitals remain below the Fermi level, then this interaction is actually slightly repulsive (because antibonding orbital combinations are always raised in energy slightly more than the corresponding bonding combinations are lowered). If the splitting is sufficient, however, the antibonding combination may be pushed above the Fermi level, whereupon its electrons drain away into surface orbitals. Clearly this gives a net decrease in energy, and

Donation implies that electrons are transferred from an adsorbate molecule to the surface. **Back-donation** refers to the transfer of electrons from the surface to the molecule. Both may contribute to covalent bonding.

acts to stabilise the adsorption bond. The net movement of electrons is from the molecule towards the surface in this case.

Type IV interactions are the inverse of the previous type, and occur when linear combinations are taken between an initially unoccupied adsorbate orbital and an initially unoccupied surface orbital. If both the resulting mixed orbitals remain above the Fermi level, then the interaction is energetically neutral, since both remain unoccupied. If the splitting is large enough, however, the bonding combination may be pushed below the Fermi level. Electrons flow from surface orbitals in the vicinity of the Fermi level into this state, decreasing the overall energy of the system and stabilising the adsorption bond. The net movement of electrons is from the surface towards the molecule.

In comparison with the ionic mode of bonding, net charge transfer in the covalent mode tends to be rather smaller, in part because each individual contribution involves only partial transfer of electrons between the bonding partners, but also because charge flow from interactions of Types II and IV often compensates for charge flow from interactions of Types I and III. Nevertheless, it is worth stressing that covalent bonding between adsorbate and surface is usually polar, to some degree.

3.5 Exercises

3.1 Confirm that the integral in Eqn. 3.7 does indeed yield the expression in Eqn. 3.8, and demonstrate explicitly that the latter tends to the appropriate limiting value as $z \to 0$.

3.2 Nitrogen dioxide (NO_2) has a permanent dipole moment of 0.316 D per molecule in the gas phase. On a certain metal surface, molecules of NO_2 adsorb vertically, with their oxygen atoms towards the substrate, at a density of 10^{-9} mol.cm^{-2}. Assuming that the molecular dipole remains unaltered, and that no charge transfer occurs between the surface and the molecules, by how much (and in what direction) would the work function of the metal change?

3.3 Scanning Tunnelling Microscopy of a Cu{111} surface reveals decaying oscillations in the electron density in the vicinity of surface defects, with a characteristic spatial period of 15 Å. Ultraviolet Photoemission Spectroscopy on the same surface detects a feature lying within a gap in the projected bulk band structure, with a maximum binding energy relative to the Fermi energy of 0.42 eV at the centre of the 1BZ. Assuming that both phenomena arise from one and the same Shockley surface state, estimate the effective mass of electrons occupying that state.

3.4 A schematic model for the ideal GaAs{111} surface is shown below, but under mildly As-rich conditions it is thought that an ordered array of top-layer Ga vacancies would form, with one missing atom from each unit cell of the type shaded. Identify all dangling bonds within such a unit cell, and predict their occupancies, assuming that electrons are able to migrate between nearest-neighbour atoms. Speculate on any likely consequences for relaxation of the remaining top-layer atoms.

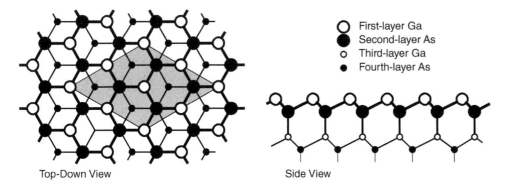

○ First-layer Ga
● Second-layer As
o Third-layer Ga
• Fourth-layer As

Top-Down View   Side View

3.5 Ethylene epoxidation is a multi-billion dollar-per-annum industrial process, but the desired reaction

$$2C_2H_4 + O_2 \rightarrow 2C_2H_4O$$

is in competition with complete combustion

$$C_2H_4 + 3O_2 \rightarrow 2CO_2 + 2H_2O$$

which is thermodynamically favoured. The consensus is that epoxidation is kinetically favoured in heterogeneous catalysis when adsorbed oxygen remains in molecular form, rather than dissociating on the surface, so the most effective catalysts are relatively inert coinage metals in preference to more reactive transition metals. Some dissociation of oxygen nevertheless occurs, however, even on the preferred silver catalyst. Adsorbed atomic chlorine is known to promote selectivity towards the desired reaction, although the overall activity of the catalyst is slightly reduced. Rationalise the role of the halogen promoter, with reference to the likely bonding modes of chlorine and oxygen with the metal surface.

3.6 **Summary**

- Reflection of bulk-like electrons from the surface sets up self-interference that leads to Friedel oscillations in the electron density, with characteristic wavelength π/k_F.
- Friedel oscillations imply the existence of a surface dipole, contributing to a potential energy barrier that retains electrons within the solid. This barrier is augmented by bulk exchange and correlation interactions.
- The minimum energy required to liberate an electron from the solid is known as the work function. Its value depends upon the particular surface facet through which the electron emerges.
- Projection of bulk band structure onto a two-dimensional plane yields a continuum of allowed energy-wavevector regions, punctuated by forbidden gaps.

- Truly surface-localised electronic states are possible only within forbidden gaps of the projected bulk band structure. As such states pass into the allowed regions, they become incompletely localised surface resonances.

- Common types of surface-localised electronic states include Shockley and Tamm varieties. Shockley states are free-electron like within the surface plane, while Tamm states are fully localised dangling bonds.

- Bonding of adsorbates may occur through physisorption or chemisorption. Physisorption arises from dipolar interactions of various types, while chemisorption arises through either ionic or covalent interactions.

- Ionic chemisorption occurs when an occupied adsorbate state naturally lies above (or when an unoccupied adsorbate state naturally lies below) the Fermi level of the surface.

- Covalent chemisorption occurs when the highest occupied and lowest unoccupied states of an adsorbate straddle the surface Fermi level.

Further reading

Argile, C., and Rhead, G. E., 1988. 'The Coadsorption of Caesium and Potassium on Ag{111}'. *Surf. Sci.* 203 (1–2): 175.

Ashcroft, N. W., and Mermin, N. D., 1976. *Solid State Physics* New York: Harcourt Brace Jovanovich.

Blyholder, G., 1964. 'Molecular Orbital View of Chemisorbed Carbon Monoxide'. *J. Phys. Chem.* 68 (10): 2772.

Crommie, M. F., Lutz, C. P., Eigler, D. M., and Heller, E. J., 1995. 'Quantum Corrals'. *Physica D* 83 (1–3): 98.

Fiete, G. A., and Heller, E. J., 2003. 'Theory of Quantum Corrals and Quantum Mirages'. *Rev. Mod. Phys.* 75 (3): 933.

Gurney, R. W., 1935. 'Theory of Electrical Double Layers in Adsorbed Films'. *Phys. Rev.* 47 (6): 479.

Hoffmann, R., 1988. 'A Chemical and Theoretical Way to Look at Bonding on Surfaces'. *Rev. Mod. Phys.* 60 (3): 601.

Lorrain, P., Corson, D. P., and Lorrain, F., 1988. *Electromagnetic Fields and Waves*, 3rd ed. New York: W. H. Freeman and Company.

Srivastava, G. P., 1997. 'Theory of Semiconductor Surface Reconstruction'. *Rep. Prog. Phys.* 60 (5): 561.

4 Kinetics and Dynamics

4.1 Introduction

The picture we have thus far painted amounts to a 'still life' of the surface. In the last chapter, the behaviour of electrons was explored subject to the implicit assumption that their environment could be considered stationary on the time-scale appropriate to their motion. In the chapter before that, we described the symmetry and structure of surfaces, including the possibility that both may change through relaxation or reconstruction, but our focus was on the start and end states, not the mechanism by which change might actually occur. Similarly, we began this book by discussing aspects of surface thermodynamics, but our interest lay with the eventual equilibrium state, not with the approach to it. Now, at last, we must turn our attention to the fundamental processes that occur at surfaces—adsorption, desorption, vibration, reaction—understanding them as ongoing events taking place *during* the passage of time, as opposed to historical facts to be explained *after* time has elapsed. In so doing, we shall address both kinetics (variation of macroscopic system properties) and dynamics (microscopic motion of atoms and molecules) to present a 'moving picture' of the surface as a place of constant agitation and continual change.

4.2 Adsorption

Whilst deriving adsorption isotherms in Section 1.7, we simply indicated that the rate of direct (as opposed to precursor-mediated) adsorption would be proportional both to the number of empty sites on the surface and to the gas-phase pressure of the adsorbing species. The constant of proportionality was largely glossed over, except to note that it would depend upon the temperature of the adsorbing gas. Now, however, we shall explicitly write the adsorption rate (adsorbates per site per second) as

$$r_a = k_a P \theta_0^n$$
$$= F \theta_0^n \kappa \exp(-E_a/RT) \qquad [4.1]$$

where P represents the gas pressure, θ_0 the relative coverage of empty surface sites, R the molar gas constant, and T the gas temperature. The parameter n determines the kinetic order of the adsorption process with respect to empty surface sites, which we expect should be first-order ($n = 1$) if species adsorb intact, or second-order ($n = 2$) if they dissociate into two fragments upon impact (reflecting the probability of the necessary number of empty sites being present at the point of impact). The rate constant, k_a, is exponentially dependent upon an activation barrier for adsorption, E_a, and the factor F simply represents the flux of adsorbates onto the surface (impacts per site per second). Finally, the parameter κ is a dimensionless entropic factor, whose value (between zero and unity) reflects the probability that the surface and adsorbate present themselves to each other in a manner that optimises adsorption. For example, an approaching adsorbate may bind to the surface significantly better when oriented in a particular way, or when impinging at a particular angle, while another may bind best if it reaches the surface at the precise moment when a particular vibrational mode reaches its maximum displacement. Grouped together, the non-flux factors are often referred to as the **sticking probability**, s, so that we may simply write

$$r_a = Fs \tag{4.2}$$

while the flux factor is generally assumed to conform to expectations from the kinetic theory of gases[1], thereby amounting to

$$F = \frac{PA_0N_A}{\sqrt{2\pi MRT}} \tag{4.3}$$

where A_0 represents the surface area per adsorption site, N_A the Avogadro constant, and M the molar mass of the adsorbing species (see Exercise 4.1).

Non-activated adsorption

When investigating the dynamics of adsorption processes, the most important point to recognise is that the potential-energy landscape navigated by an incoming adsorbate is inherently multi-dimensional. Even in the simplest possible case—that of an atomic adsorbate—the system possesses three degrees of freedom corresponding to the adsorbate's position in space relative to the surface, and multiple additional degrees of freedom associated with possible displacement of surface atoms. For a molecular adsorbate, there will be up to three further degrees of freedom connected with its orientation, not to mention any degrees of freedom linked with internal deformations. A complete description of the adsorption process would require not only that we keep track of the system's position within this multidimensional potential-energy landscape, but also its multidimensional velocity across that landscape. Unsurprisingly, the temptation is to simplify, in the interests of gaining qualitative understanding, by ignoring inconvenient degrees of freedom wherever possible.

Adopting the most extreme stance, we might claim that only one degree of freedom is fundamentally important; namely, the adsorbate's distance from the surface. In taking this position, we neglect any internal and orientational degrees of freedom possessed by the adsorbate (natural if the adsorbate is only a single atom but questionable in all other cases) and similarly neglect any degrees of

[1] Any suitable text may be consulted; for example, Atkins et al., 2017.

freedom relating to the adsorbate's lateral position across the surface, not to mention those internal degrees of freedom connected with the surface itself. The benefit of such an approach, however, is that we can represent the potential-energy profile within a one-dimensional approximation, as depicted in Fig. 4.1. Here, the surface–adsorbate interaction creates a potential well in the vicinity of the surface (of depth equal to the heat of adsorption, q_a) that asymptotically levels off as the adsorbate departs from the surface, but rises rather steeply when the adsorbate approaches the surface too closely.[2] We shall assume a surface temperature such that an adsorbate thermalised in the potential well has insufficient energy to readily escape, implying a significant lifetime in its bound state. The question nevertheless remains as to how the adsorbate enters the well in the first place.

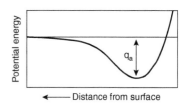

Fig. 4.1 Non-activated adsorption, modelled with a single degree of freedom. The surface lies towards the right, vacuum towards the left.

Let us consider, therefore, the implications of this one-dimensional adsorption model in forensic detail. Coming from the gas phase, the potential energy of the adsorbate is initially only weakly dependent upon distance from the surface, while its kinetic energy will be similarly almost constant. As its trajectory brings it closer to the surface, however, the adsorbate's potential energy will drop—gradually at first, but ever more rapidly—and the kinetic energy will correspondingly increase. In effect, the adsorbate will accelerate towards the surface, until it passes the minimum of the potential well, whereupon it will rapidly decelerate before coming to a halt. From this turning point, the adsorbate will now precisely retrace its recent trajectory in reverse, eventually departing from the surface carrying exactly the same kinetic energy that it originally possessed before the interaction took place. The result is simply an example of elastic scattering, and demonstrates that adsorption is strictly impossible in a system that really does possess only a single degree of freedom. If we wish to insist upon a one-dimensional potential-energy landscape, we must at least implicitly accept that other degrees of freedom exist. Adsorption can only occur if, in the course of the adsorbate's interaction with the surface, energy is lost from the single explicit degree of freedom by transference into these others. This may imply energy transfer from the adsorbate to the surface, but may equally involve the excitation of adsorbate vibrational modes (either internal molecular modes or frustrated rotational and/or translational modes). Importantly, the propensity of the adsorbate to adsorb in the case depicted in Fig. 4.1 is not dependent upon it having any particular amount of kinetic energy when it first approaches the surface, meaning that this is a non-activated process.

Turning to a more complex example, let us imagine a system where a single degree of freedom would be entirely insufficient to fully describe the process. Just such a case may be found in dissociative adsorption, where one of the adsorbate's internal bonds is broken while the bond to the surface is formed. We shall describe this situation with two degrees of freedom—one corresponding to the distance of the adsorbate from the surface and the other to the length of the breaking bond. A typical potential-energy landscape for such a system is depicted

---

[2] For the purposes of the present discussion, we need not trouble ourselves as to the particular origin of the surface–adsorbate potential well, which may arise due to either physical or chemical forces, as discussed in Section 3.4.

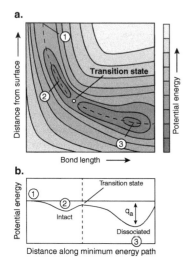

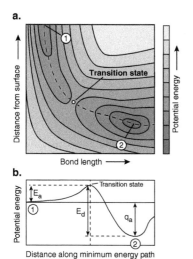

Fig. 4.2 Non-activated dissociative adsorption, modelled with two degrees of freedom (cf. Fig. 4.1).

When a system travels through a potential-energy landscape along its minimum-energy path, the distance along that path can be regarded as the **reaction coordinate** for the motion.

Fig. 4.3 Activated dissociative adsorption (cf. Fig. 4.2).

in Fig. 4.2a. When the adsorbate is far from the surface, there is a significant potential energy cost associated with breaking the bond, and when the bond is broken on the surface there is a significant potential energy cost associated with the desorption of the products. Between these two extremes, the landscape features a characteristic elbow-shaped channel through which the minimum-energy path for dissociative adsorption passes. It remains possible (Fig. 4.2b) to plot a one-dimensional energy profile, by following the potential energy along this minimum-energy path, but the generalised **reaction coordinate** representing distance along the path no longer corresponds to a single natural degree of freedom, but rather to an ever-varying combination of two.

In the particular example shown, this kind of plot reveals the useful information that although the minimum-energy path contains a local maximum (transition state) this occurs at a potential energy below that possessed by the adsorbate in the gas phase. As the adsorbate approaches the surface, therefore, two likely scenarios may play out. In the first of these, energy loss from the two explicit degrees of freedom may be relatively slow, so that the transition state is reached with sufficient remaining kinetic energy to surmount it with ease. In this case, the adsorbate is likely to enter the deeper potential well of the dissociated state before thermalising. In the second scenario, energy loss from the two explicit degrees of freedom may be more rapid, so that the total energy of the adsorbate falls below that of the transition state before it is reached. In this case, the adsorbate will thermalise whilst in the shallower potential well of the intact state. The relative energy between the transition state and the gas phase ensures, however, that subsequent entry into the dissociated state will be exponentially more probable than desorption. In either scenario, the likelihood is that any adsorbate entering the double well would eventually find itself localised in the deeper of the two, regardless of how much kinetic energy it started with in the gas phase. Once again, therefore, this is a non-activated process overall. In common with all such cases, the activation barrier, E_a, appearing in Eqn. 4.1, may simply be set to zero. A corollary of this is that the effective activation barrier for *desorption*, which we may reasonably label E_d, is simply equal to the overall heat of adsorption, q_a.

Activated adsorption

In the preceding example, the transition state in the minimum-energy path for dissociative adsorption was found to lie below the potential energy of the gas-phase adsorbate, leading to a non-activated process. There is no particular reason, however, why this transition state must necessarily be so situated, and it may indeed sometimes be found to lie *above* the potential energy of the gas-phase adsorbate. In such cases, the overall process of adsorption must be an activated one, depending upon sufficient kinetic energy being possessed by the molecule before it interacts with the surface. If energy transfer from the explicit degrees of freedom to the implicit ones is slow, on the time-scale of the adsorbate's approach to the transition state, then the critical kinetic energy required will be equal to E_a, the activation barrier for adsorption (Fig. 4.3). The activation barrier for desorption, E_d, will correspondingly be equal to the sum $q_a + E_a$.

Notwithstanding the energetic requirement for the adsorbate to overcome an activation barrier for adsorption, the two-dimensional nature of the process implies a further consideration, which will have a bearing upon the κ factor in Eqn. 4.1. Specifically, it will be important to ask if the adsorbate's total kinetic energy (translational plus vibrational) can be efficiently coupled to the reaction coordinate in the vicinity of the transition state. The answer will depend upon whether the transition state in the minimum-energy path occurs at an early or a late stage in the adsorption process.

Consider a molecule in its vibrational ground state, initially following a trajectory closely aligned with the reaction coordinate. If the transition state occurs quite early along the minimum-energy path, increasing the translational energy will naturally assist in accessing additional adsorptive trajectories passing very close to the transition state (Fig. 4.4a). In contrast, if the transition state occurs rather later along the minimum energy path, increasing the molecule's translational energy will enhance adsorption only somewhat inefficiently, through trajectories that do not pass particularly close to the transition state (Fig. 4.4b).

Alternatively, rather than increasing the adsorbate's translational energy, one could instead increase its vibrational energy, in an attempt to facilitate bond cleavage. In the case of an early transition state, however, any mismatch between the

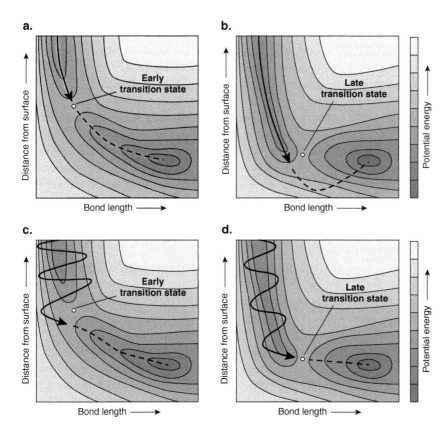

Fig. 4.4 Early and late transitions states, showing trajectories influenced by addition of translational or vibrational energy.

vibrational phase and the adsorbate's moment of arrival may result in a trajectory that does not pass close to the transition state (Fig. 4.4c). At best, the addition of vibrational energy in such a scenario will be a very inefficient way of enhancing adsorption, and could conceivably even hinder it. With a late transition state, the situation is again reversed. Although synchrony between the adsorbate's vibrational phase and its moment of arrival is once again crucial, it is now possible that trajectories may be accessed that pass considerably closer to the transition state than would ever have been possible for a molecule in its vibrational ground state (Fig. 4.4d). Addition of vibrational energy may therefore be a rather efficient strategy for enhancing adsorption when the transition state occurs late in the process.

Trapping and steering

In view of the entropic factor alluded to above, it ought to be clear that possession of adequate kinetic energy alone is not sufficient to ensure passage of an incoming molecule across an adsorption activation barrier. Neither does it follow that molecules with higher kinetic energy will necessarily always enjoy a higher sticking probability than those with less, although they often do. Indeed, careful kinetic studies with supersonic molecular beams (Section 5.7) reveal a few cases where sticking probability actually decreases with increasing translational energy, albeit only in the very low-energy regime (Fig. 4.5). Two different mechanisms have typically been invoked to explain this phenomenon, depending upon whether evidence suggests that the eventual adsorbed molecule retains any 'memory' of its initial state or not. Both explanations hinge upon the notion that the transition state for adsorption is relatively inaccessible, and that a molecule may not find itself passing close to it without either some form of guidance or the opportunity to make multiple attempts. It can often be difficult, in practice, to distinguish between the two phenomena.

The mechanism known as **steering** recognises that any incoming molecule will be guided towards the transition state by the shape of the potential-energy landscape through which it moves. Clearly, slower-moving molecules will have their trajectories altered rather more strongly than those arriving with greater translational kinetic energy (Figs. 4.6a and 4.6b). Accordingly, a slow-moving molecule may have only just enough kinetic energy available to surmount the activation barrier, but it does have the best opportunity for that kinetic energy to be steered into the appropriate reaction coordinate, passing rather accurately through the transition state itself. A fast-moving molecule, on the other hand, may have plenty of kinetic energy available, but is more likely to miss the precise transition state, and hence to need most or all of that excess energy in order to traverse a reaction path that is not quite the lowest energy that might have been found by a slower molecule. For sufficiently fast molecules, the availability of excess energy typically dominates, and the sticking probability exhibits the expected rise as the translational kinetic energy increases, but in the very low translational energy regime it may be that steering implies a reversal of that trend.

An alternative to steering is the mechanism known as **trapping**, in which the possibility of direct adsorption from the gas phase is set aside. Here, the incoming

Sticking probability in activated adsorption may be enhanced due to **steering** (where an adsorbate is guided through the transition state by the potential-energy landscape) or **trapping** (where an adsorbate makes many attempts to cross the transition state whilst weakly bound to the surface).

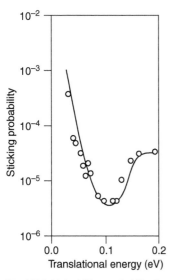

Fig. 4.5 Initial sticking probability for methane adsorption on Pt{110}(1 × 2) (adapted from Walker and King, 2000).

molecule does not succeed in immediately surmounting the activation barrier for adsorption, but instead becomes trapped in a weakly bound precursor state. The trapped state may either be physisorbed or weakly chemisorbed, but in either case the molecule becomes thermalised and will remain on the surface with some relatively brief lifetime. In that period, there exists a certain probability that the molecule may pass through the transition state into a more strongly adsorbed state, and it may have sufficient time to make a very large number of attempts, but the successful passage will likely be long after any memory of the molecule's incoming state has faded. In the regime where trapping is important, therefore, the overall sticking probability will vary according to the likelihood that the incoming molecule becomes thermalised in the precursor state, rather than simply bouncing straight back off the surface on its initial impact. Unsurprisingly, slow-moving molecules are more likely to be trapped (Figs. 4.6c and 4.6d) and so once again a reversal of the expected trend in sticking probability may arise, with lower translational energies linked to higher adsorption rates.

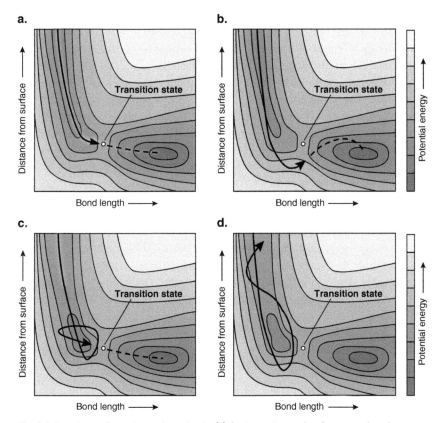

Fig. 4.6 Steering and trapping trajectories. In (a) the incoming molecule approaches the surface slowly and is steered through the transition state, whereas in (b) the molecule approaches more rapidly, failing to pass precisely through the transition state. In (c) the incoming molecule approaches the surface slowly and becomes trapped in a shallow potential well before traversing the transition state, whereas in (d) the molecule approaches more rapidly and fails to become trapped, leaving the surface without sticking.

As in the case of steering, however, sufficiently high translational energies will eventually facilitate direct adsorption without the need for trapping, and the more usual increase of sticking probability with increasing translational energy will typically reassert itself in that regime.

4.3 Desorption

We have already dealt briefly with desorption of surface species in Section 1.7, where our focus was very much on its role in establishing the defining equilibrium underlying adsorption isotherms. In the present section, we shall devote our attention to the kinetics of desorption in rather more detail.

Temperature-Programmed Desorption (TPD)

One of the most fundamental questions faced by a surface scientist in studying a new system is over what range of conditions might adsorbed species be expected to remain present on the surface. In order to establish this basic fact, it is common to start out by performing TPD experiments, which typically take the following form. Having first prepared a clean surface under ultra-high vacuum, the species of interest is adsorbed (usually by exposure to its vapour) at a sufficiently low surface temperature that an appreciable coverage is built up. Then, after once more fully evacuating the experimental chamber, the surface is heated at a constant rate (typically $1\text{-}10\text{K.s}^{-1}$) whilst monitoring whatever species may desorb by means of a suitably located mass spectrometer.[3] The mass spectrometer signal for a particular species may ideally be taken as a proxy for the desorption rate of that species, and analysed accordingly.

It is important to mention, in passing, that there is scope for ambiguity in the interpretation of mass spectrometry data, even for quite simple molecular species. Ultra-high vacuum chambers are usually equipped only with quite basic spectrometer models, which are not always capable of high resolution in both time and mass, so it is common to record data for integer molecular mass numbers only (as opposed to actual molecular masses). Unfortunately, this means that some commonplace species may readily be confused with one another. Molecular nitrogen, for example, has a mass number of 28, but so too does carbon monoxide, assuming the most common isotopes are involved in both cases.[4] In order to distinguish between such species, it may be necessary to employ isotopically labelled gases, or to examine in detail the **cracking pattern** of

When molecules are ionised upon entry into a mass spectrometer, some of them may dissociate into smaller fragments that show up in the spectrum at lower masses. Fortunately, the array of fragments forms a predictable **cracking pattern** for a given parent species, and so can be used to aid identification.

---

[3] The inlet of the mass spectrometer should be close to the sample, and in direct line-of-sight, to avoid the possibility that desorbing species may be intercepted before entering the instrument. In practice, the location is often not critical, but when particularly 'sticky' species (e.g. water, ammonia, etc.) are involved, a poorly sited mass spectrometer may delay, broaden, and/or differentially attenuate the corresponding signal.

[4] Molecules of $^{12}C^{16}O$ and $^{14}N_2$ have molecular masses of 27.995 and 28.006 atomic mass units, respectively, and thus could be distinguished unequivocally by a sufficiently precise mass spectrometer.

the expected species. On entry into a mass spectrometer, ionisation of the initially neutral species stimulates a certain percentage of the incoming entities to fragment, leading to a characteristic pattern of signals at lower mass than that of the parent molecule. Thus, to continue with the same example of molecular nitrogen, its cracking pattern is known to contain a signal at mass 14 (from isolated nitrogen atoms generated by dissociation in the mass spectrometer) in addition to the expected signal at mass 28. Needless to say, the signal at mass 14 is *not* expected from the cracking pattern of carbon monoxide, which should include signals at masses 12 and 16 instead. Careful consideration of published cracking patterns, including the expected relative intensities, can be used to unequivocally identify species in many cases, albeit the analysis is not always straightforward when multiple different species may be desorbing from the surface at the same time.

Zero-, first-, and second-order desorption

At the outset, it is worth recognising that kinetic analysis will be much simplified if we assume that the parameters controlling desorption are independent of relative coverage (as per our previous derivation of adsorption isotherms). In reality, variation of desorption parameters may alter the behaviours described here, rendering the analysis unreliable or even impossible, so the following cases ought to be viewed as idealised archetypes rather than immutable categories.

We begin with a general expression (the **Polanyi–Wigner equation**) for the desorption rate (adsorbates per site per second) in the form

$$r_d = k_d \theta^n$$
$$= v\theta^n \exp(-E_d/RT) \qquad [4.4]$$

The **Polanyi–Wigner equation** predicts the desorption rate based upon relative coverage, surface temperature, the desorption barrier, and a pre-exponential factor.

where θ represents the relative coverage, R the molar gas constant, and T the surface temperature. The rate constant k_d (glossed over in Section 1.7) is exponentially dependent upon the activation barrier for desorption, E_d, and also involves a pre-exponential factor, v, that may be thought of as an attempt frequency for desorption. The parameter n simply dictates the kinetic order of the desorption process, and it is this aspect that we shall now explore.

Turning first to the case of **zero-order desorption** ($n = 0$) our expression for the rate simplifies to the form

$$r_d = k_d$$
$$= v\exp(-E_d/RT) \qquad [4.5]$$

In **zero-order desorption** (as occurs when there is a reservoir of sub-surface adsorbates, or when there are multilayers) the rate does not depend upon the relative coverage.

from which we may expect the desorption to accelerate as the surface temperature is linearly increased over time. A plot of the rate versus surface temperature will therefore take the form shown in Fig. 4.7, rising rapidly, before abruptly crashing back to zero when the supply of adsorbed entities (atoms or molecules) suddenly runs out. In fact, the figure depicts the results of several separate TPD experiments, conducted with a series of different initial relative coverages. The technique is quantitative, in the sense that the area under each

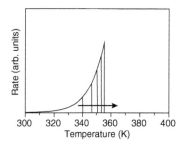

Fig. 4.7 TPD traces for zero-order desorption, starting from a range of initial relative coverages (0.2, 0.4, 0.6, 0.8 and 1.0) increasing as indicated by the arrow.

curve ought to be proportional to the initial coverage, and two important features may be discerned: firstly, traces obtained with different initial coverages all share a common leading edge; and secondly, the temperature at which the rate collapses shifts higher as the initial coverage is increased. These diagnostic features constitute the signature of zero-order desorption. Such a situation arises, in practice, when the supply of adsorbed entities amenable to desorption does not vary over time until all are gone, perhaps because we are depleting a multi-layer (where each departing entity merely reveals an identical entity beneath) or because we are depleting a sub-surface reservoir (where each entity leaving the surface is replaced by an identical entity diffusing up from the bulk). Either way, the zero-order behaviour should come to a (sudden) halt only once the multi-layer or sub-surface reservoir is fully depleted.

If, on the other hand, the stock of independently desorbing entities is *not* continually replenished, then we may reasonably suppose the rate of desorption to be directly proportional to their relative coverage. That is, we should anticipate

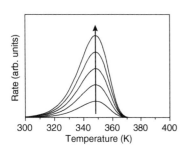

Fig. 4.8 TPD traces for first-order desorption, starting from a range of initial relative coverages (0.2, 0.4, 0.6, 0.8 and 1.0) increasing as indicated by the arrow.

In **first-order desorption** (as occurs when adsorbed species depart intact from a single surface layer) the rate is proportional to the relative coverage.

$$r_d = k_d\theta$$
$$= v\theta\exp(-E_d/RT) \tag{4.6}$$

which corresponds to **first-order desorption** ($n = 1$). In these circumstances, the acceleration in rate as the surface temperature increases is counterbalanced by a deceleration in rate as the relative coverage is progressively depleted. The result expected of a TPD experiment, depicted in Fig. 4.8, is an asymmetric peak when plotting desorption rate against surface temperature. Diagnostically, we note that the temperature at which the rate is maximised remains constant when the initial coverage is varied, and that neither the leading nor the trailing edge of the trace is common from one experiment to another. Such behaviour is most commonly observed in the desorption of entities whose adsorbed and gas-phase forms are essentially alike (e.g. molecules that adsorb intact from the gas phase).

Finally, we consider the case where desorption requires recombination of two identical adsorbed fragments (e.g. individual atoms recombining to desorb as homonuclear diatomic molecules) so that the process depends upon the probability of two such fragments occupying adjacent sites. This leads to the case of **second-order desorption** ($n = 2$) characterised by

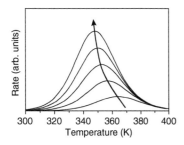

Fig. 4.9 TPD traces for second-order desorption, starting from a range of initial relative coverages (0.2, 0.4, 0.6, 0.8 and 1.0) increasing as indicated by the arrow.

In **second-order desorption** (as occurs when two identical fragments recombine to form the departing species) the rate is proportional to the square of the relative coverage.

$$r_d = k_d\theta^2$$
$$= v\theta^2\exp(-E_d/RT) \tag{4.7}$$

and depicted in Fig. 4.9. In this case, the TPD trace obtained when starting with one specific initial coverage is more nearly symmetric than in either of the previous two cases, but it is notable that the temperature at which the rate is maximised shifts progressively lower as the initial coverage is increased, and that the trailing edges of traces obtained with differing initial coverages are common.

Redhead formula and beyond

Having established, by means of the qualitative diagnostic features described above, that a particular desorption process occurs with zero-, first-, or second-order kinetics, it is natural to wish for a means by which the crucial quantitative parameters, v and E_d, may be obtained from the available data.[5] As a start, we may try differentiating the desorption rate (Eqn. 4.4) with respect to temperature, and setting the result equal to zero, yielding

$$\frac{v}{\beta}\exp\left(\frac{-E_d}{RT_m}\right) = \frac{E_d}{RT_m^2} \tag{4.8}$$

in the first-order case, where T_m is the temperature at which the rate is maximised, and where β represents the heating rate (see Exercise 4.2). Since both of these quantities are known, this equation links the values of our two unknown parameters, v and E_d, but does not uniquely provide the value of either. Nevertheless, it is common practice to assume that the pre-exponential factor, v, takes a value in the ballpark of 10^{13} s^{-1}, as this is a typical attempt frequency for bond-breaking processes in other areas of chemistry. The desorption activation barrier, E_d, may then be calculated, albeit the equation requires one to do so iteratively, by refining a succession of guesses, since it cannot be rewritten in closed form. Fortunately, the value for E_d obtained by this method is only relatively weakly dependent upon the exact value assumed for v, although an error of several orders of magnitude in the latter parameter could still shift the apparent value of the former parameter by a not entirely negligible amount.

In order to avoid the inconvenience of an iterative procedure, surface scientists often resort to using an approximation to Eqn. 4.8, stated as

$$E_d = RT_m\left[\ln\left(\frac{vT_m}{\beta}\right) - 3.64\right] \tag{4.9}$$

and known as the **Redhead formula** (for first-order desorption). Its validity was asserted by Redhead (1962) for v/β in the range $10^8 - 10^{13}$K^{-1}, which is sufficiently broad to encompass most likely scenarios.[6] Furthermore, one may deduce that the error in estimating E_d amounts to approximately ($T_m/50$) kJ.mol^{-1} for every order of magnitude error in one's estimate of v.

Notwithstanding the ubiquity of the Redhead formula, a variety of approaches have nevertheless been proposed to circumvent our initial ignorance concerning the value of the pre-exponential factor, just one of which (also suggested by Redhead) will be described here. The essence of the technique is to conduct a series of TPD experiments at different heating rates, β, and to analyse how this affects the temperature at which the maximum desorption rate occurs.

Starting with Eqn. 4.8, for first-order desorption, one may rearrange to obtain

Subject to a reasonable guess for the pre-exponential barrier, the **Redhead formula** provides an estimate of the desorption barrier, based upon knowledge of the temperature at which desorption is maximised for a given heating rate.

$$\ln\left(\frac{RT_m^2}{\beta}\right) = \left(\frac{E_d}{R}\right)\frac{1}{T_m} + \ln\left(\frac{E_d}{v}\right) \tag{4.10}$$

---

[5] Note that analysis of TPD data provides insight into the desorption barrier, not the adsorption heat. The two quantities are equal only in the absence of any activation barrier for adsorption. This is often indeed true, but should not be carelessly assumed.

[6] The Redhead formula is sometimes quoted with 3.64 replaced by 3.46, apparently in error. The practical effect of this discrepancy is generally rather negligible.

whereupon it becomes clear that a plot of $\ln(RT_m^2/\beta)$ versus $1/T_m$ should yield a straight line with gradient E_d/R and intercept $\ln(E_d/v)$. Measurement of both gradient and intercept will, therefore, permit both E_d and v to be separately extracted from the data. Whilst experiments performed in this manner confirm that an estimate of $10^{13}\,\text{s}^{-1}$ for the pre-exponential factor is often quite reasonable, it is clear that some variability in this parameter should be expected from case to case, and that naive application of the Redhead formula with an estimated value of v is not always certain to produce reliable results for the desorption activation barrier (see Exercise 4.3). The additional effort required to repeat TPD experiments at a variety of heating rates is certainly justified if accurate measurements of desorption parameters are desired.

Finally, it is important to remember that the foregoing discussion implicitly assumes both the desorption activation barrier and the pre-exponential factor to be independent of surface coverage. Clearly, this is unlikely to be strictly true for any real system, since lateral interactions affecting the heat of adsorption (Section 1.8) are equally likely to influence these desorption parameters. It may not, therefore, always be possible unequivocally to identify the kinetic order of desorption without a pre-existing theoretical model, inclusive of coverage effects, against which to compare experimental data.

4.4 Vibration

Having described the various means by which an adsorbate may either arrive at or depart from the surface, we will shortly turn our attention towards their dynamic behaviour (intramolecular and frustrated modes) during whatever time they spend there. Before doing so, however, it will prove both instructive and interesting to enquire into the vibrational properties of the underlying surface itself.

Phonon modes

It is beyond the scope of this book to provide a comprehensive introduction to the topic of vibrational excitations in crystals, but here let us simply note that imposition of periodic boundary conditions on harmonic atomic motion suggests a wavelike treatment not dissimilar to the Bloch description of electronic wavefunctions.[7] Complete knowledge of a crystal's classical vibrational behaviour thus amounts to obtaining a set of eigenvalues (vibrational frequencies) and their corresponding eigenfunctions (atomic displacement amplitudes) at each point within the system's first Brillouin zone (1BZ). By analogy with our discussion of electronic structure (Section 3.3) we may imagine projecting the bulk vibrational bands onto a surface-parallel plane, anticipating that features lying outside of the projected continuum will reveal surface-localised vibrational states worthy of our interest. Before investigating further, however, it will be

---

[7] For a detailed treatment, see Srivastava, 2022.

advisable to consider some matters of quantisation, polarisation, and polarity, which set vibrational band structures apart from electronic ones.

Regarding quantisation, in the electronic case it turns out that the number of available electronic states with a given wavevector is essentially infinite (if we include unbound states) but that the number of quanta associated with each state (the number of electrons it contains) is restricted to values of zero, one, or two. Furthermore, the total number of quanta within the 1BZ (the total number of electrons) is fixed by the charge neutrality of the system as a whole. In contrast, the number of available vibrational states with a given wavevector is fixed by the number of atoms per primitive unit cell (three times more states than atoms) and it is the number of quanta (in each such state) that is unbounded. This last point reflects the fact that whilst electrons (the fundamental quanta of electronic excitation) obey Fermi–Dirac statistics, **phonons** (by which name we shall now refer to the fundamental quanta of vibrational excitation) obey Bose-Einstein statistics.

The second important distinction between phonons and electrons is that the underlying vibrational modes of the former may be classified into **longitudinal** and **transverse** categories, depending upon whether the atomic displacements lie parallel or perpendicular to the wavevector. In isotropic or continuous media, precisely one third of the modes will be purely longitudinal in nature, while the remainder will be purely transverse. In anisotropic media, longitudinal and transverse modes will generally mix, unless the wavevector lies along a sufficiently high-symmetry direction.

Our final observation, before turning to surface-specific considerations, must be the categorisation of vibrations into **acoustic** and **optical** modes, according to whether atoms within a single primitive unit cell move in phase with one another or not. The description of the out-of-phase modes as 'optical' references the fact that these are the only modes that can be infrared active (if the vibrating atoms carry dissimilar charges). A solid will always possess precisely three acoustic modes, while the remainder will all be optical. Characteristically, the frequencies of all three acoustic modes fall linearly to zero in the zero-wavenumber (long-wavelength) limit. Optical modes, in contrast, have finite frequency but zero gradient (with respect to wavenumber) in that same limit. Since the speed of a phonon (we state without proof) is proportional to the gradient of its underlying vibrational frequency, it follows that the speed of long-wavelength sound within a crystal is dominated by the acoustic modes (hence their name).

Concerning surfaces, much of the early groundwork (pardon the pun) was laid by researchers interested in the propagation of seismic waves. Since the length-scales involved were decidedly macroscopic, the underlying bulk medium was invariably approximated as an isotropic continuum. Nevertheless, a number of important principles were established, not least a recognition that the inherent anisotropy of the surface region means categorisation of waves as longitudinal or transverse tends to be of limited utility. It is generally more insightful to speak of displacements occurring either within the **sagittal plane** (containing both the wavevector and the surface-normal vector) or along the **shear horizontal** direction (perpendicular to the sagittal plane).

Phonons are quantised vibrational excitations in a liquid or solid material. They may be classified as either **longitudinal** or **transverse** in nature, depending upon the direction in which atoms move relative to the wavevector of the excitation.

Acoustic phonons involve only in-phase motion within each primitive unit cell, while **optical** phonons involve some out-of-phase motion. Only the latter may couple with electromagnetic radiation.

Surface phonons are often classified according to their **sagittal plane** or **shear horizontal** character, rather than as purely longitudinal or transverse.

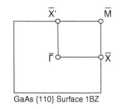

GaAs {110} Surface 1BZ

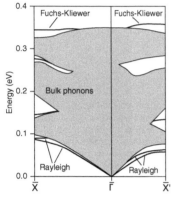

Fig. 4.10 Phonon band structure of GaAs{110}. Projected bulk bands are shown as shaded regions, while isolated lines show surface-localised modes (adapted from Fritsch et al., 1993).

Rayleigh and **Love modes** are surface acoustic phonons, typically found at frequencies below the lowest bulk modes.

Fuchs–Kliewer modes are surface optical phonons, typically found at frequencies above the highest bulk modes.

Softening refers to a drop in frequency of a phonon mode as ambient conditions are varied, indicating the potential onset of a phase transition (such as a surface reconstruction).

Furthermore, Lord Rayleigh (1885) demonstrated that the surface of an isotropic continuum must necessarily support a surface-localised acoustic mode (the **Rayleigh mode**) polarised within the sagittal plane and lying lower in frequency than any of the bulk modes (Fig. 4.10). Somewhat later, Love (1911) showed that a surface-localised acoustic mode of shear horizontal polarisation (the **Love mode**) may also exist at similarly low frequencies. In the context of crystalline solids, where the isotropic continuum model breaks down, modes reminiscent of Rayleigh and Love nevertheless exist, although generally of mixed polarisation except at high-symmetry wavevectors. The distinction between Rayleigh and Love modes is thus somewhat diluted, and the literature tends to refer to all acoustic surface modes lying beneath the projected bulk modes by the name of Rayleigh (proving that Love does not, alas, conquer all).

Turning to optical modes, surface-localised versions may (and often do) turn up within the internal gaps of the projected bulk band structure, but a thorough discussion of these would be excessive here.[8] We should, however, take a moment to note the existence of one very particular optical mode lying above the upper edge of the bulk states at the surfaces of polar materials. This is the so-called **Fuchs–Kliewer mode**, distinguished by such strong coupling to infrared radiation that it ought properly to be regarded (at least in the long-wavelength limit) as a polariton, which is to say a composite quasiparticle comprising inseparable oscillations of both vibrational and electromagnetic character (Fuchs and Kliewer, 1965). Although not particularly strongly surface-localised, compared with other surface optical modes, its evanescent infrared activity makes it especially noteworthy.

One further interesting aspect of surface phonon modes might usefully be highlighted at this point, which is that their atomic displacement patterns may be regarded as attempts at surface reconstruction that result in increased surface free energy and hence are ultimately disfavoured. The phonon frequency is thus a measure of how unstable the corresponding reconstruction would have been, and particularly low phonon frequencies are indicative of potential reconstructions that could perhaps occur under only slightly different conditions. For instance, a given surface phonon mode, at a particular wavevector, may be observed dropping in frequency (or **softening**) as the temperature is increased, finally reaching zero when the corresponding reconstruction becomes favourable from the perspective of surface free energy. In this way, the softening of surface phonons may be considered diagnostic of incipient reconstructions in systems that do not yet satisfy the necessary conditions for a permanent change actually to occur.

Molecular modes

Turning now to vibrations of adsorbates attached to solid surfaces, let us begin with the case of an isolated adsorbed molecule. Such a molecule would, of

---

[8] The interested reader is referred to the recent book on surface dynamics by Benedek and Toennies (2018).

course, have exhibited a well-defined set of vibrational frequencies whilst in the gas phase, and the majority of these will carry over to the surface context in a one-to-one manner. For those molecular modes with frequencies far above the maximum phonon frequency of the surface,[9] the main consequence of adsorption will simply be that electronic effects may strengthen or weaken certain intramolecular bonds and hence shift the normal mode frequencies. For example, in the case of CO adsorption, back-donation into the $2\pi^*$ orbitals (Section 3.4) is likely to weaken the C–O bond, and the corresponding vibrational frequency is typically red-shifted as a result. Indeed, in this particular case, the degree of red-shift in different environments is so well documented that it can be taken as something of a proxy for the adsorption heat, and even used to infer the local bonding of the adsorbate. That is to say, a red-shift for adsorbed CO of around $100\,cm^{-1}$ is indicative of atop binding, around $200\,cm^{-1}$ suggestive of bridge binding, and around $300\,cm^{-1}$ or more is usually linked with binding into a hollow site. For other adsorbate species, the phenomenology is rarely so developed, but nevertheless the information can be very useful in hinting at the mode of adsorption. Note, however, that not all molecular modes will be red-shifted upon adsorption. A good example of the opposite behaviour may be found in the case of ammonia adsorption, where formation of a dative adsorption bond typically promotes withdrawal of electrons from the molecule and a notable blue-shift of the NH_3 umbrella mode.

At the other extreme, let us consider not an isolated molecule on the surface, but instead an ordered overlayer, in which molecules are packed sufficiently tightly together that they interact strongly with one another. Here, our expectation must be that intermolecular coupling will not only influence vibrational frequencies, but also enforce a description in terms of two-dimensional phonons. In the long-wavelength limit, molecules in neighbouring primitive unit cells will execute vibrational motion in perfect phase with one another, while at the shortest wavelength compatible with lattice periodicity these same molecules will vibrate in exact antiphase. These two cases will typically exhibit extremal frequencies between which the frequencies of mid-wavelength oscillations will be found. In other words, the vibrational frequencies of individual molecules broaden into a two-dimensional vibrational band structure. This point is rather important because the predominant measurement technique for observation of molecular vibrations at surfaces, namely infrared spectroscopy, is sensitive only to long-wavelength oscillations (Section 5.8). In the case of an ordered adsorbate overlayer, therefore, it will measure only one extremal frequency, corresponding to the infinite-wavelength limit, rather than the full range.

Finally, we might enquire as to the behaviour of a disordered array of molecules on the surface, close enough to one another that we might anticipate strong vibrational coupling, but lacking the periodicity necessary to justify a phonon-based description. In this connection, we note the phenomenon of Anderson localisation (Anderson, 1958) in which strong disorder within a crystal

---

[9] The maximum phonon frequency in most solids is typically no more than a few hundred wavenumbers. In systems with particularly stiff bonds (e.g. diamond, graphene, etc.) it may exceed $1000\,cm^{-1}$, but these are very much the exception rather than the rule.

tends to impede the transport of quantised excitations, localising them on individual sites. Here, the upshot is that quantised vibrational excitations (phonons) cannot travel across the surface, instead being trapped on individual molecules. Since the phase relationship between different sites will now be essentially random, infrared spectroscopy ought not to be particularly sensitive to any one vibrational state over others, and our expectation should be that any measurable variation in vibrational frequencies must reflect heterogeneity amongst local environments, rather than coupling between sites.

Frustrated modes

Above, we have considered vibrations within the material comprising the surface itself, and vibrations within molecules adsorbed upon it. Now, we shall consider vibrational modes in which an adsorbate moves relative to the surface, usually referred to as **frustrated translations** and **frustrated rotations**.

The simplest frustrated translational mode (indeed the only one guaranteed to be present for *any* adsorbate) is that associated with motion perpendicular to the surface. Whether travelling towards or away from the surface at any given moment, it is clear that a turning point must eventually be reached if the adsorbate is not either to penetrate beneath the surface or to desorb from it. If this mode is sufficiently excited, desorption or in-diffusion will occur, but so long as the adsorbate remains at the surface, free translation is frustrated and vibrational motion ensues (Fig. 4.11a).

In addition to the perpendicular mode just described, frustrated translations may also be found in the two surface-parallel dimensions (Fig. 4.11b/c). In general, corrugation of the lateral potential-energy landscape will imply certain preferred adsorption sites for any given adsorbate, and if the available thermal energy is sufficiently low then individual adsorbates will be trapped within individual potential-energy wells. Lateral vibrations within these wells

Frustrated translations and **frustrated rotations** relate to degrees of freedom that would be non-vibrational for species in the gas phase, but which gain an oscillatory character when that species is adsorbed on a surface. The frequencies of such modes are invariably very low, compared with intramolecular vibrations.

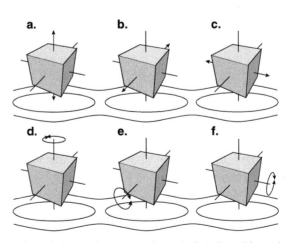

Fig. 4.11 Frustrated translations and rotations schematically indicated for molecules (represented as cubes) localised in discrete adsorption sites.

then constitute frustrated translations. When more thermal energy is available, however, it may be that one or both of these frustrated translations give way to genuine diffusion across the surface. Similarly, we note the existence of up to three frustrated rotational modes (Fig. 4.11d/e/f) that may exist for molecular adsorbates—two relating to rotation about surface-parallel axes, and a third relating to rotation about the surface normal itself. Once again, with sufficiently low thermal energy, molecular adsorbates will settle into preferred orientational potential-energy wells, oscillating within these according to well-defined frustrated rotations. With more thermal energy, some or all of these modes may give way to genuine rotational motion, albeit with periodic variation of angular speed.

All of the frustrated modes, it ought to be mentioned, typically occur at rather low frequencies, often around $100 \, cm^{-1}$ or below. As such, they are readily distinguishable from most intramolecular vibrational modes, but they may be strongly coupled with the phonon modes of the surface. Ultimately, it is the flow of energy between phonons and the frustrated modes that is responsible for thermal reorientation and/or diffusion of molecules at surfaces.

4.5 Reaction

If frustrated translations and rotations may be considered as failed attempts at molecular reorientation or surface diffusion, inter- or intra-molecular vibrations might similarly be viewed as failed attempts at chemical reaction. How, then, does interaction with the surface influence the dynamics and kinetics of chemical reactivity?

The Brønsted–Evans–Polanyi (BEP) relation

To begin, let us imagine a particular chemical reaction, occuring in the absence of a nearby surface, with some well-defined reaction coordinate leading smoothly from reactants to products across a transition state (see Fig. 4.12). The forward rate will be proportional to a rate constant, k_r, that varies exponentially with the activation barrier for the reaction, E_r, and linearly with some pre-exponential factor, v, that depends upon the entropy difference between initial and transition states. That is, we have

$$k_r = v \exp(-E_r/RT) \tag{4.11}$$

with R and T the molar gas constant and system temperature respectively, as above (cf. Sections 4.2 and 4.3). All other conditions being equal, it is clear that whilst variation in the pre-exponential factor may modulate the rate constant to some degree, the activation barrier is by far the dominant parameter in determining the reaction rate.

Now consider the very same reaction taking place with reactants, products, and transition state all bound to a surface. The absolute energies of all three states are likely to be substantially shifted, but we shall focus only upon their

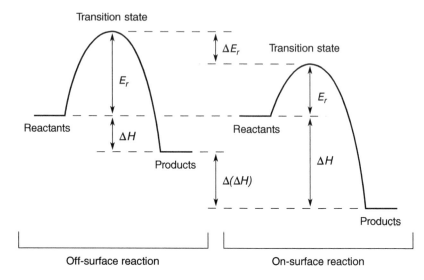

Fig. 4.12 Gas-phase and surface reaction barriers. Energies of the reactant states have been aligned for clarity, but in reality may differ considerably.

The **Brønsted–Evans–Polanyi (BEP) relation** empirically links a change in reaction enthalpy to a change in activation barrier, when comparing a given type of reaction across different local environments.

relative dispositions. In this regard, the presence of the surface may alter the overall reaction enthalpy, ΔH, and also the activation barrier, E_r (Fig. 4.12). Any change in the former will have consequences for the final equilibrium ratio between reactants and products, but only a change in the latter will alter the forward reaction rate. Interestingly, however, there is often a linear empirical relationship—the **Brønsted–Evans–Polanyi (BEP) relation**—between these two important energies, which may be expressed as

$$\Delta E_r \propto \Delta(\Delta H) \qquad\qquad [4.12]$$

where $\Delta(\Delta H)$ indicates a change in the reaction enthalpy, and ΔE_r a corresponding change in the activation barrier.[10] It follows, from this relationship, that knowledge of the activation barrier corresponding to some *particular* value of the reaction enthalpy may be used (if the constant of proportionality in Eqn. 4.12 is also known) to predict the activation barrier in situations where the reaction enthalpy is different. The constant of proportionality is likely to be close to zero if the transition state occurs early along the reaction path (with a configuration similar to the initial state) and close to unity if the transition state occurs late (with a configuration similar to the final state).

To give a concrete example, consider the case of methane adsorption on transition metals, where calculations across a range of surfaces indicate an activation energy for the first dehydrogenation step ($CH_4 \rightarrow CH_3 + H$) that varies

[10] Why an approximately linear relationship between reaction enthalpy and activation energy might be expected to hold is explained rather clearly in the original paper by Evans and Polanyi (1938), to which the curious reader is referred.

linearly with changes in the reaction enthalpy, as shown in Fig. 4.13.[11] Since, in this specific case, the initial state features a gas-phase molecule, while both the transition state and final state are bound to the surface, it is no surprise that the slope of the graph is close to unity. Once sufficient data has been assembled to determine both the slope and the intercept with some confidence, the activation energy for dehydrogenation on a yet-to-be-studied surface may be estimated simply by measuring or calculating the corresponding dissociative adsorption heat (this being minus the reaction enthalpy in the present example) and reading from the graph. As a general principle, one may reasonably infer that the activation energy for a particular mechanistic step in a sequence of surface reactions is likely to be much reduced when interaction with the surface increases the exothermicity (or reduces the endothermicity) of that step, and vice versa.

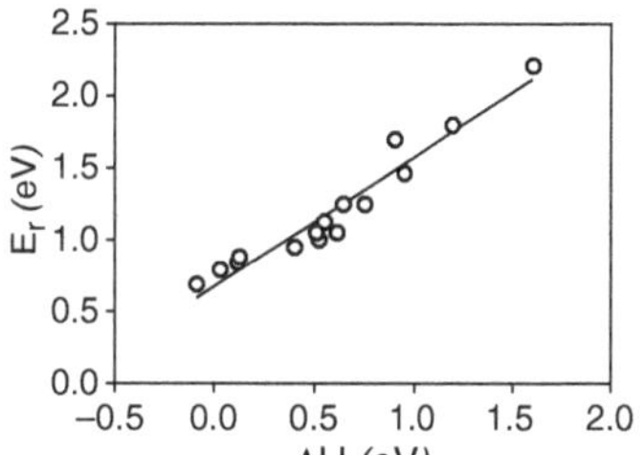

Fig. 4.13 Calculated BEP relation between activation energy, E_r, and reaction enthalpy, ΔH, for the first dehydrogenation of methane on a variety of transition metal surfaces (after Nørskov et al., 2008).

The Sabatier principle and Volcano plots

Building upon the notion of the BEP relation, and in particular its implications for the activation energies of dissociation reactions taking place at surfaces, we may draw some quite profound conclusions regarding the suitability of different materials as heterogeneous catalysts. Taking ammonia synthesis ($N_2 + 3H_2 \rightarrow 2NH_3$) via the Haber–Bosch process, for example, where dissociation of molecular N_2 is generally considered to be at least partially rate-limiting, the BEP concept allows us to anticipate that the activation barrier for this mechanistic step will be smallest when the reaction takes place on a surface that binds the product N atoms particularly strongly, and largest when the surface binds N atoms only rather weakly. Pleasingly, then, it turns out that transition metals from the left-hand side of the d-block, empirically found to have high heats for dissociative adsorption of dinitrogen, are also found to have some of the highest sticking probabilities for the same process. Unfortunately, these metals are *not* very efficient Haber–Bosch catalysts, because their surfaces also bind ammonia very strongly, and hence they rapidly become self-poisoned with a product molecule that readily forms on the surface but desorbs only very reluctantly.

In fact, assuming that the rate of N_2 dissociation becomes exponentially faster as one moves from right to left across the periodic table, but that the rate of NH_3 desorption becomes exponentially slower over the same transition, then there will inevitably exist a sweet spot for ammonia production when neither one nor the other process utterly dominates the overall rate (Fig. 4.14). The very best pure metals for Haber–Bosch catalysis are probably ruthenium and osmium, but both are far too expensive and toxic to be viable for a process carried out on an industrial scale. Cheap and safe, iron is the practical compromise.

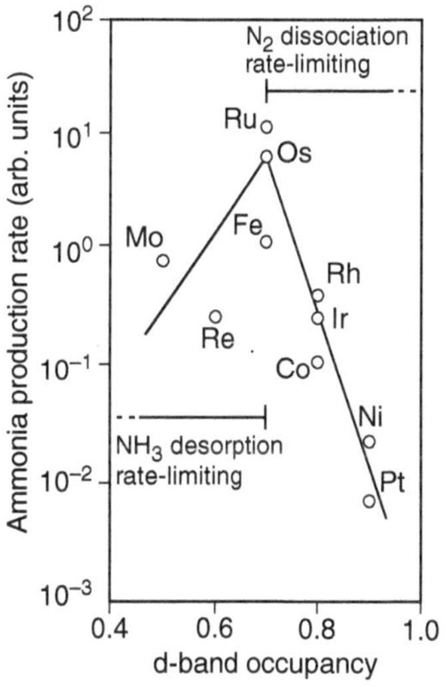

Fig. 4.14 Volcano plot for the rate of ammonia production over different transition-metal catalysts (after Ozaki and Aika, 1981).

[11] In this particular example, the reaction enthalpy is often found to be positive, which is to say that the initial step in the dissociative adsorption of methane is usually endothermic. Subsequent dehydrogenation steps, leading to adsorbed methylene, methylidine. or carbon, may be either exothermic or endothermic, and BEP relations of differing slope could be sought for each of these in turn.

Dissociation is, indeed, thought to be the rate-determining step in the mechanism for this catalyst, but not to such a degree that removal of ammonia can be entirely ignored under industrial conditions. Any potential replacement for iron as the Haber–Bosch catalyst of choice would need to shift the reaction further towards the ideal point of balance between nitrogen dissociation and ammonia desorption, without being unfeasibly costly or toxic—a combination of qualities that has largely resisted over a century of research into the topic.

Graphs of the kind shown in Fig. 4.14 occur frequently in catalysis, whenever the reaction in question requires competing imperatives to be balanced by a compromise catalytic candidate (the **Sabatier principle**). For obvious reasons, these key diagrams are known as **volcano plots**, and they represent an important starting point for screening possible catalysts prior to evaluation.

Langmuir–Hinshelwood vs. Eley–Rideal

For surface reactions that involve association of reactants, rather than dissociation, the obvious question we must address is how the involved entities actually approach one another, and here two alternative models will necessarily dominate the discussion. The **Eley–Rideal mechanism** proposes that one of the reacting entities is already adsorbed on the surface prior to the reaction step itself, but that the second approaches it directly from the gas phase. The **Langmuir–Hinshelwood mechanism**, in contrast, posits that *both* entities are pre-adsorbed before the reaction step, approaching each other by a process of surface diffusion.

The attraction of the Eley–Rideal model is that the adsorption heat liberated as the gas-phase entity approaches the surface, expressed initially as translational energy, may be channelled into the reaction coordinate before it is dissipated into surface vibrational modes. In the Langmuir–Hinshelwood model, on the other hand, both adsorbed entities are thermalised with the surface and neither will possess any very significant kinetic energy. From the viewpoint of surmounting the reaction activation barrier, therefore, the Eley–Rideal concept seems eminently more promising than the Langmuir–Hinshelwood scenario. That said, however, the Eley–Rideal mechanism does hinge upon the incoming entity impinging directly upon the pre-adsorbed entity in its first approach to the surface, in the correct orientation and internal configuration to facilitate reaction, which involves a high degree of coincidence and is thus entropically disfavoured. Although the Langmuir–Hinshelwood mechanism undoubtedly involves a lower probability of surmounting the activation barrier on any *single* attempt, it does benefit from the fact that two adjacent pre-adsorbed entities can make repeated reaction attempts during the lifetime of their sojourn as neighbours. From a practical perspective, the preferred mechanism for a particular reaction may readily be inferred from analysis of reaction kinetics (see Exercises 4.4 and 4.5). Over the decades since they were first proposed, the Langmuir–Hinshelwood mechanism has proven to be applicable in a substantial majority of important heterogeneous catalytic reactions, but the Eley–Rideal mechanism may nevertheless apply in certain very specific cases.

The **Sabatier principle** embodies the notion that catalytic activity is greatest when the reactant(s) bind to the catalyst neither too weakly to instigate chemical change nor too strongly to imply slow desorption of the product(s).

Volcano plots display a peak in reaction rate when two competing rate-determining steps are balanced.

The **Eley–Rideal mechanism** for surface reactions involves an adsorbed species interacting with a species impinging directly from the gas phase. Contrast this with the **Langmuir–Hinshelwood mechanism**, in which both species are adsorbed before the reaction begins.

4.6 Exercises

4.1 An initially clean surface is exposed to carbon monoxide at a partial pressure of 10^{-10} mbar (10^{-8} Pa) and a temperature of 100 K. The initial sticking probability is measured to be 0.5, and the surface density of adsorption sites is $10^{15}\,cm^{-2}$. Assuming that adsorption is non-dissociative and Langmuirian, and that desorption is negligible, how long would it take to reach half of saturation coverage? Compare with pressures of 10^{-7}, 10^{-2}, and 10^{3} mbar.

4.2 Differentiate Eqn. 4.6 with respect to temperature. Set the result to zero, and hence obtain Eqn. 4.8 for the temperature at which the first-order desorption rate is maximised. Why is this procedure somewhat less useful in the zero-order and second-order cases?

4.3 Given the following data for first-order desorption at different heating rates, use the Redhead formula (Eqn. 4.9) to estimate the activation barrier for desorption (assuming some reasonable pre-exponential factor) for each heating rate separately. Then, use Eqn. 4.10 to obtain both the activation barrier *and* the pre-exponential factor, by plotting the data for different heating rates on a suitable graph. Comment on the discrepancy between the two methods.

$\beta\,(K.s^{-1})$	0.5	1.0	2.0	5.0	10.0	15.0	25.0
$T_m\,(K)$	450	455	463	475	485	490	495

4.4 The surface-catalysed reaction $A+B\rightarrow C$ proceeds via a Langmuir–Hinshelwood mechanism. Assuming that the surface coverages of all species follow the form given in Eqn. 1.43 (with non-dissociative adsorption throughout) write a general expression for the rate of forward reaction in terms of their gas-phase pressures, P_A, P_B, and P_C. Comment upon the kinetics in circumstances where (i) one of the reactants adsorbs much more strongly than either the product or the other reactant; and (ii) the product adsorbs much more strongly than either reactant.

4.5 How would your answers to the last question differ if the reaction proceeded via an Eley–Rideal mechanism?

4.7 Summary

- The rate of adsorption is most generally described as the product of incident flux with a sticking probability that depends exponentially upon an activation barrier, linearly upon a pre-exponential factor relating to the entropy of the adsorption transition state, and also upon the fraction of empty surface sites raised to some power.

- A full description of adsorption dynamics should be multidimensional, but one- and two-dimensional approximations may yet provide useful insights if treated with caution.

- In non-activated adsorption, sticking occurs independently of the incoming adsorbate's kinetic energy. An entropic factor may nevertheless limit the sticking probability, due to steric considerations.

- In activated adsorption, there exists a threshold incident kinetic energy beyond which the sticking probability rapidly increases. The distribution of energy between different degrees of freedom may be important, and the time available for energy transfer can be crucial. Trapping and steering may be relevant processes in the low kinetic energy regime.
- The rate of desorption depends exponentially upon an activation barrier, linearly upon a pre-exponential factor that may be regarded as an attempt frequency, and also upon some power of the relative coverage.
- The kinetic order of desorption may be established by analysis of peaks obtained in a TPD experiment, and especially by consideration of whether the temperature at which the desorption rate is maximised rises, falls, or stays the same when the initial coverage is increased.
- An individual first-order TPD peak may be analysed to obtain the activation barrier for desorption (e.g. by the Redhead method) in which case the pre-exponential factor must be guessed, but a family of such peaks measured with different heating rates may be analysed exactly without the necessity of such a guess.
- Quantised surface-localised vibrations (surface phonons) of various types may be observed at frequencies disallowed in the bulk (cf. surface-localised electronic states).
- Intramolecular vibrations persist at surfaces, with frequencies either blue- or red-shifted due to electronic modulation of intramolecular bond strengths concomitant with adsorption.
- Frustrated translations and rotations couple strongly with surface phonons, and the exchange of energy amongst these modes is largely responsible for surface reorientation and/or diffusion of adsorbates.
- The Brønsted–Evans–Polanyi relation provides a link between reaction activation barriers and reaction enthalpies, with implications for the volcano plots used when evaluating potential catalyst candidates.

Further reading

Anderson, P. W., 1958. 'Absence of Diffusion in Certain Random Lattices'. *Phys. Rev.* 109 (5): 1492.

Atkins, P., de Paula, J., and, Keeler, J., 2017. *Atkins' Physical Chemistry*, 11th ed. Oxford: Oxford University Press.

Benedek, G., and Toennies, J. P., 2018. *Atomic Scale Dynamics at Surfaces* Berlin: Springer.

Evans, M. G., and Polanyi, M., 1938. 'Inertia and Driving Force of Chemical Reactions'. *Trans. Faraday Soc.* 34 (25): 11.

Fritsch, J., Pavone, P., and Schröder, U., 1993. 'Ab Initio Calculation of Surface Phonons in GaAs(110)'. *Phys. Rev. Lett.* 71 (25): 4194.

Fuchs, R., and Kliewer, K. L., 1965. 'Optical Modes of Vibration in an Ionic Crystal Slab'. *Phys. Rev.* 140 (6A): A2076.

Love, A. E. H., 1911. *Some Problems of Geodynamics* Cambridge: Cambridge University Press.

Nørskov, J. K., Bligaard, T., Hvolbaek, B., Abild-Pedersen, F., Chorkendorff, I., and Christensen, C. H., 2008. 'The Nature of the Active Site in Heterogeneous Transition Metal Catalysis'. *Chem. Soc. Rev.* 37 (10): 2163.

Ozaki, A., and Aika, K., 1981. 'Catalytic Activation of Dinitrogen'. In *Catalysis: Science and Technology*, Vol. I, edited by John R. Anderson and Michel Boudart, Berlin: Springer-Verlag.

Rayleigh, Lord (Strutt, J. W.), 1885. 'On Waves Propagated along the Plane Surface of an Elastic Solid'. *Proc. London Math. Soc.* s1-17 (1): 4.

Redhead, P. A., 1962. 'Thermal Desorption of Gases'. *Vacuum* 12 (4): 203.

Srivastava, G. P., 2022. *The Physics of Phonons*, 2nd ed. Boca Raton: CRC Press.

Walker, A. V., and King, D. A., 2000. 'Dynamics of Dissociative Methane Adsorption on Metals: CH_4 on Pt{110}(1×2)'. *J. Chem. Phys.* 112 (10): 4739.

Techniques

5.1 Introduction

In this chapter we review a variety of specialised techniques commonly applied to surface systems. Most are routinely conducted in ultra-high vacuum environments, for reasons we shall address shortly, but a few methods suitable for ambient-pressure gaseous or liquid environments are also included. An overview of first-principles computational theory in surface science rounds out the discussion.

The importance of ultra-high vacuum conditions for surface science techniques is essentially two-fold. First, there is the issue of attenuation, whenever a particular technique relies upon the passage through space of photons, electrons, or a molecular beam. This is equally true whether the particles in question are impinging upon the surface or emanating from it; ultra-high vacuum drastically extends the relevant mean free path. The second reason relates to the necessity of obtaining a clean surface with which to work. Although surfaces outside of the laboratory are likely covered in a variety of different species, and may host copious sub-surface impurities, the complexity of such a situation precludes any hope of understanding the system in atomistic detail. Surface scientists often aim to build a bottom-up understanding by starting with the simplest possible model for the real-world scenario, adding complexity in stages along the way. Producing a clean surface is the essential prerequisite for this strategy, and ultra-high vacuum conditions prove to be critical in this endeavour.

Consider, therefore, the surface of a freshly prepared sample. However pure the material is purported to be, one should note that the total number of impurity atoms within a macroscopic sample will vastly exceed the number needed to establish a saturated layer on the surface. Indeed, the undercoordinated (and hence reactive) nature of the surface often tends to encourage migration of impurities from the bulk. Fortunately, removal of species from the surface itself may be achieved relatively easily, by some combination of (i) heating, if the adsorbed species is sufficiently weakly bound; (ii) sputtering, through the physical impact of electrically accelerated noble gas ions; or (iii) titration, exposing to a gas-phase species known to react with the impurity to yield a weakly bound product. Less fortunately, the most generally effective of these methods—sputtering—tends also to disorder the surface, so a subsequent period of annealing is usually necessary if this tactic is employed, albeit this inevitably aids in the migration of sub-surface impurities to replenish the stock in the surface layer.[1]

Only by repeated cycles of heating/sputtering/titration and annealing can a situation be achieved whereby a region adjacent to the surface (but many layers deep) has at length become denuded of impurities altogether. Experiments may now reproducibly be performed, without risk of impurities 'bubbling-up' from below, so long as sample temperatures are maintained beneath the threshold for bulk impurity mobility. This initial preparation phase may take anything from a few days to several weeks, depending upon the properties of the bulk material. Subsequent cleaning of the surface between experiments ought only to require brief sputtering or mild titration, and annealing should also be relatively brief and mild. Consideration of the adsorption rate equation (Section 4.2) allows one to estimate, however, that the surface would hypothetically remain clean for barely a nanosecond under ambient pressure conditions in a moderately re-active atmosphere. In the ultra-high vacuum regime ($c.10^{-10}$ mbar) the window within which the surface remains clean is stretched to several hours.

5.2 Electron diffraction techniques

The use of X-Ray Diffraction (XRD) in establishing the structure of bulk crystals is ubiquitous, and covered in most textbooks dealing with the solid state. At sur-faces, however, the use of X-rays for this purpose is extremely limited, because they typically penetrate deep into the material and are hence only marginally influenced by the region we wish to study.[2] In consequence, diffraction exper-iments in surface science are generally conducted with entities that penetrate at most a few layers into the studied material, and foremost amongst these are electrons, either at low energy and near-normal incidence (discussed below) or at higher energy and grazing incidence (not described here). We might mention, in passing, that helium diffraction experiments also rely upon somewhat similar concepts, but are not covered in this book. It ought to be stressed, however, that all diffraction techniques ultimately hinge upon crystalline order, and none have much to contribute to the study of disordered surfaces.

Low-Energy Electron Diffraction (LEED)

In a typical LEED experiment, electrons with energies in the approximate range 10–1000 eV impinge upon the surface at normal incidence.[3] At these energies, they penetrate only a few nanometres into the sample, with the

---

[1] We distinguish here between 'heating' the surface briefly, to encourage desorption of adsorbed species, and 'annealing' the surface at length, to encourage its return to crystallinity from a state of disorder.

[2] Grazing Incidence X-ray Diffraction (GIXD) achieves surface sensitivity through the phenom-enon of total external reflection, but this introduces complications and limitations of its own.

[3] LEED experiments at off-normal incidence are perfectly possible, although the analysis is a little more involved. A discussion of the general case may be found in Jenkins (2020).

a.

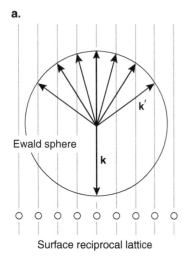

Ewald sphere

k

k′

Surface reciprocal lattice

b.

Orthographic projection of spots

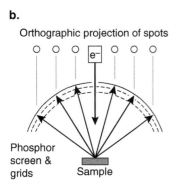

e−

Phosphor
screen &
grids   Sample

Fig. 5.1 Reciprocal and real-space geometry of a LEED experiment, showing (a) the Ewald sphere construction, and (b) the arrangement of electron gun, sample, grids, and screen.

majority being elastically scattered back into the region from whence they came. The electron diffraction patterns arising in the process may best be understood with reference to the Ewald sphere construction, shown in Fig. 5.1a. Conservation of energy constrains the wavevectors ($\mathbf{k}'$) of scattered electrons to lie with their tips touching a sphere of radius equal to the wavenumber, $|\mathbf{k}|$, of the incoming electrons, while conservation of momentum constrains the surface-parallel component of the outgoing wavevectors to match one of the reciprocal lattice vectors of the surface.[4] Together, these constraints dictate a limited number of discrete directions along which outgoing electrons must be scattered.

LEED experiments are generally conducted using a hemispherical phosphor-coated screen whose shape echoes the Ewald sphere, in order to simplify analysis (Fig. 5.1b). Between the sample and the screen, two or three metal grids are usually interposed. The first of these is held at a potential chosen to retard approaching electrons, so as to prevent the passage of inelastically scattered (and hence lower energy) electrons to the screen. The function of the other grid (or grids) is to accelerate the remaining elastically scattered electrons towards the screen, so as to maximise the glow produced upon impact. Conveniently, an orthographic projection of the spots created when these electrons eventually hit the screen should match precisely the reciprocal lattice of the surface, scaled by a factor related to the electron energy.[5] The technique thus provides a very direct method of assessing the size and symmetry of the surface's two-dimensional lattice (see Exercise 5.1). Changes due to surface reconstruction or overlayer adsorption are particularly clearly detected by this approach, and for this reason the examination of LEED patterns is routinely included in basic protocols for determining the surface cleanliness and orderliness of single-crystal samples.

Further information may also be gleaned through analysis of variations in the spot intensity as a function of electron energy. By systematically varying the potential difference through which the impinging electrons are accelerated (V) and measuring the intensity of an individual spot (I), the experimentalist compiles a series of so-called I/V curves for as many spots as can be observed over an appreciable energy range. In the analysis of XRD data for bulk crystals, it is possible directly to invert spot intensity data to determine atomic positions within the unit cell, but only because one may reasonably assume each photon scatters no more than once during its interaction with the material. In electron diffraction, limited penetration into the sample (which is, after all, essential for the surface sensitivity of the technique) implies

---

[4] It is a consequence of Bloch's theorem (Section 3.3) that states with wavevectors lying outside of the first Brillouin zone (1BZ) can be equally well represented with wavevectors that have been modified to lie inside the 1BZ by the addition of a reciprocal lattice vector. Wavevectors at crystalline surfaces are, therefore, only uniquely defined to within the addition of an arbitrary surface reciprocal lattice vector.

[5] As the electron energy increases (decreases) so the Ewald sphere expands (shrinks) and the angles at which spots are observed become smaller (larger) so that the diffraction spots eventually converge towards the centre of the screen (diverge beyond the edge of the screen).

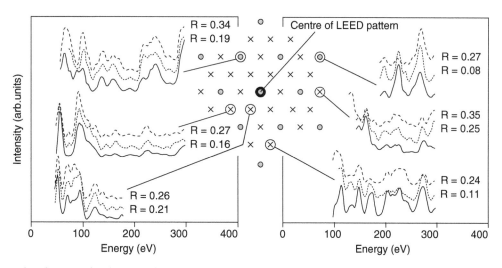

Fig. 5.2 Examples of measured and simulated LEED-I/V curves upon coadsorption of O and CO on Ru{0001}. Solid curves represent experimental data for various diffraction spots in the six-fold symmetric LEED pattern. Spots depicted schematically as filled circles are present when the surface is clean; those depicted as crosses appear only after adsorption. Dashed and dotted curves are simulations in which CO molecules are, respectively, constrained to be upright or permitted to tilt. R-factors for individual spots can be combined to give overall values of 0.33 and 0.17 for upright and tilted molecules (after Narloch et al., 1995).

that electrons likely scatter several times before leaving. Direct inversion of LEED-I/V data is not, therefore, generally considered feasible. Instead, standard practice is to computationally simulate diffraction for several competing structural models of the surface. A figure of merit, usually referred to as an R-factor, is then calculated for each structure, to express how closely or otherwise the simulated I/V curves match the experimental ones, a value of zero implying perfect correlation, while a value of unity implies no correlation at all (Fig. 5.2). The R-factor then provides an objective basis upon which to systematically refine the most promising structural models. If only a single model can be found for which the R-factor may be driven much below about 0.2, then it is usually reasonable to view this as representing a more-or-less definitive structure for the surface.

5.3 Scanning probe techniques

Amongst the most widely used techniques in surface science, scanning probe microscopies are most often deployed to produce stunning images of surface morphology at atomic or near-atomic resolution (Fig. 5.3), but they may also be used in less straightforward modes to provide considerably more than just visual immediacy. Despite key similarities, however, it is essential to distinguish two distinct branches of microscopy according to the phenomenon (current or force) that comprises the surface probe. We shall consider the two possibilities separately below.

a.

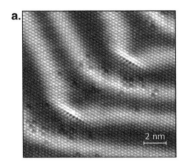

2 nm

b.

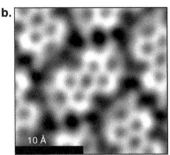

10 Å

Fig. 5.3 (a) STM image of Au{111} after low-coverage adsorption of Pd. Clearly visible is evidence of the classic 'herringbone' reconstruction, in which the top-layer of atoms becomes rumpled due to an increase in its atom density relative to the bulk. Individual atoms may nevertheless be discerned; scattered anomalous features are identified as substitutional Pd atoms (after Baber et al., 2010); (b) AFM image showing an array of PTCDA (3,5,9,10-perylene-tetracarboxylic-dianhydride) adsorbed on Ag{111} (Mönig et al., 2018).

Scanning Tunnelling Microscopy (STM)

In an STM experiment, a very sharp metal tip is held around a nanometre from the sample surface, and a small bias voltage, V, is applied to drive a current of electrons either from tip to sample or from sample to tip. Crucially, the gap between tip and sample provides a potential barrier across which the electrons must tunnel, and this quantum mechanical phenomenon turns out to be exponentially dependent upon the tip–sample separation.[6] Indeed, if the barrier were to have a rectangular potential profile, of height τ and width s, then the tunnel current would be proportional to $\exp(-2s(2m_e\tau/\hbar^2)^{1/2})$ according to standard quantum-mechanical reasoning, with m_e the mass of the electron.[7] The barrier is not, of course, quite so simple in form, but as shown in Fig. 5.4 it may reasonably be approximated as such. For an electron tunnelling at energy ε above the sample Fermi level, we may take

$$\tau \approx (eV + \phi_s + \phi_t - 2\varepsilon)/2 \qquad [5.1]$$

as the effective barrier height, where, ϕ_t and ϕ_s are the tip and sample work functions, respectively, and e the elementary electronic charge.

In general, one must imagine the total current as the integral over all such elastic tunnelling processes for electrons whose energies fall within the range between the sample and tip Fermi levels, each weighted by the corresponding tip and sample densities of states. For very small bias voltages, however, only

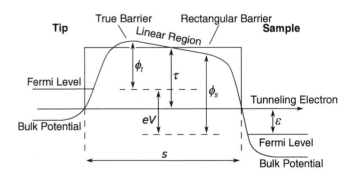

Fig. 5.4 Schematic showing the relationship between bias voltage, V, work functions, ϕ_s and ϕ_t, and the effective barrier, τ, for tunnelling in STM at an electron energy ε above the sample Fermi level.

---

[6] The very highest resolution STM experiments tend to be those conducted under ultra-high vacuum conditions, in which case the gap between tip and sample may be regarded as genuinely empty. It should be pointed out, however, that the technique may be applied quite successfully (albeit generally with less fidelity) in gaseous or even liquid environments.

[7] Note that the probability of tunnelling depends upon the barrier's height, width, and shape. This differs from the classical probability of surmounting the barrier through thermal excitation, which depends exponentially upon only the barrier's maximum height (and the prevailing temperature). The classical current would therefore display Arrhenius-type behaviour, but the quantum current does not.

tunnelling from one Fermi level to the other remains possible, and the expression for the tunnel current, I, simplifies to

$$I \propto D_F^s D_F^t \exp\left(-2s\sqrt{m_e(\phi_s + \phi_t)/\hbar^2}\right)$$ [5.2]

where D_F^t and D_F^s are the tip and sample densities of states evaluated at their respective Fermi levels.

The precise value of D_F^t is unknown, but assumed independent of the tip position. If we further assume that the local value of D_F^s does not vary strongly from one point on the surface to another, then the exponential factor is seen to dominate the behaviour of the tunnel current, and this factor is exquisitely sensitive to the tip–sample separation (synonymous with the barrier width). Accordingly, when the tip is laterally scanned across the surface at constant height, variations in the surface topography may be recorded by monitoring the corresponding variations in tunnel current. Alternatively, a feedback loop may be employed to maintain a constant tunnel current by varying the tip height during scanning; in this mode the necessary height variations mimic the surface topography.[8] An important caveat in all of this, however, is that D_F^s can *sometimes* vary quite strongly across the surface, particularly in the locality of surface vacancies, steps or adsorbates. When this occurs, a significant electronic effect can mask or mimic topographic features. Furthermore, local changes in the surface work function, ϕ_s, can similarly occasion ambiguity in the intepretation of STM images. Note also that on semiconducting samples, images obtained when using opposing sign in the bias voltage effectively probe separately the valence and conduction bands of the material, potentially revealing the presence of occupied or unoccupied dangling bonds (see Tamm states, Section 3.3).

Notwithstanding these caveats, STM is a vital and incisive tool in efforts to understand surface phenomena, bringing with it a crucial element of real-space information that is absent from many other techniques. Nevertheless, a lack of chemical specificity constitutes a real impediment to its practical utility. Ultimately, STM images provide a view of the surface in which molecules often appear as little more than blobs. When sub-molecular resolution is achieved, these blobs may be identifiable as particular chemical species, but when the resolution is not so fine it is usually only possible to infer their identities by cross-referencing with other experimental results for the same system. That said, two forms of spectroscopy based upon the STM technique offer hope in this regard.

In the first approach, usually referred to simply as Scanning Tunnelling Spectroscopy (STS), the first derivative of tunnel current with respect to bias voltage is recorded (and usually normalised against the ratio of the same two variables). The resulting quantity, plotted as a function of bias voltage, is expected to mimic the sample density of states to some degree, although theoretical justification for this interpretation relies upon somewhat questionable assumptions. If one is prepared to accept these, however, the technique provides a valuable insight

[8] Piezoelectric actuators confer the necessary fine control over three-dimensional tip position to permit both lateral scanning and vertical adjustment on sub-Ångstrom length scales.

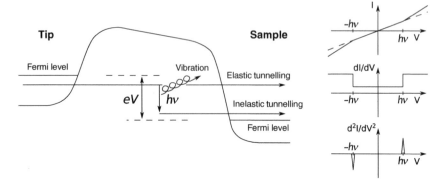

Fig. 5.5 Schematic showing how the opening of an inelastic tunnelling channel, when *eV* exceeds the vibrational quantum *hν*, leads to a discontinuous change in the gradient of the tunnel current, and hence to peaks in its second derivative with respect to bias voltage.

into local electronic structure, which may in turn aid in the identification of chemical species.

In the second approach, referred to as Inelastic Electron Tunnelling Spectroscopy (IETS), it is the second derivative of tunnel current that is generally recorded. Plotted against the bias voltage, this quantity should vary relatively smoothly (albeit usually with significant noise) except when the bias voltage passes through energies associated with quantised vibrations at the surface. At these energies, new channels for inelastic tunnelling become available (since the electron can then lose energy to a vibrational mode during its passage between tip and surface) and a peak in the second derivative is generated (Fig. 5.5). Once again, this can prove diagnostic of the species giving rise to individual features within an STM image.

Atomic Force Microscopy (AFM)

One significant limitation of the STM technique is that it works only when the sample is electrically conductive, since the tunnel current would otherwise flow only transiently, limited by charge accumulation on the sample. Metallic samples present no difficulty on this score, but semiconductors may not be suitable for imaging at low temperatures, and insulators cannot reliably be imaged at all. These issues do not, however, present an obstacle to the AFM approach. The technique is also, arguably, more tolerant of operation under ambient conditions than is STM.

AFM can be conducted in **contact mode**, **non-contact mode**, and **tapping mode**, each with distinct advantages and disadvantages.

In an AFM experiment, the sample is laterally scanned by a sharp tip, in much the same manner as described above. Rather than measuring a tunnel current between tip and sample, however, the morphology of the surface is now deduced from aspects of the force acting upon the tip, and there are three alternative ways in which this is typically achieved, namely **contact mode**, **non-contact mode** and **tapping mode**. In all three modes, the tip is mounted on a

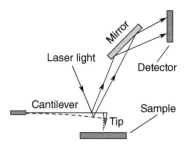

Fig. 5.6 Optical lever measurement of cantilever deflection. Very much not to scale, but showing the principle of magnification.

flexible cantilever, whose spring-like deflection may be monitored remotely (e.g. by means of an optical lever, see Fig. 5.6).

Let us begin by imagining a tip brought progressively closer to the sample, preliminary to the commencement of lateral scanning. During initial approach, the tip experiences an attractive force towards the surface (indicating the onset of physisorption) which is manifest in a measurable deflection of the cantilever. If the tip is brought sufficiently close to the surface, however, the force it experiences will eventually change sign, due to Pauli repulsion, and it is under these repulsive conditions that the cantilever deflection is most sensitive to surface topography. The simplest imaginable experiment, therefore, involves bringing the tip into this repulsive regime (the contact regime) and then maintaining constant cantilever deflection (through use of a feedback loop) whilst scanning the tip laterally across the surface; the necessary feedback signal may then be plotted to form an image of the surface based on constant repulsive force. Inevitably, however, this contact-mode imaging does carry the drawback that the tip may cause physical damage through lateral shear as it is dragged across the surface.

Non-contact mode, in which the tip remains always within the regime of attractive interaction, carries substantially less risk of sample damage, because the forces involved are considerably smaller than when operating in contact mode. Experiments of this type are generally carried out by imposing a driven vertical oscillation on the cantilever and observing variations in its dynamic motion, rather than its static deflection, within either an amplitude modulation or a frequency modulation approach. Tapping mode, as the name suggests, similarly involves driven vertical oscillation of the cantilever, but with the tip encroaching briefly into the contact regime at the closest approach of each cycle. Although the maximum force exerted by the tip upon the sample is rather larger than in either of the other two imaging modes, the lack of any lateral shear component limits the potential for surface damage.

In addition to these measurements, which provide insight primarily into the topography of the surface, further information can be provided by measuring the phase lag between the signal driving the oscillation of the cantilever and its actual motion. The latter is delayed due to the tip-surface interaction, and images whose contrast is derived from phase variation across the surface thus carry information about the elastic properties of the material. As such, they may be invaluable when assessing the mechanical properties of surfaces, offering insight at a length-scale that no competing technique can address.

Finally, it is worth noting that the force responsible for deflection of the cantilever need not be limited to simple physisorption of the tip. Variants of the technique include Magnetic Force Microscopy (MFM), in which a magnetised tip is used to image surface magnetic domains; Kelvin Probe Force Microscopy (KPFM), in which the capacitance between tip and sample is analysed to image the local work function; and AFM InfraRed (AFM-IR) spectroscopy, in which local heating of the sample by absorption of infrared laser light induces a measurable force on the tip, enabling the acquisition of localised vibrational spectra.

5.4 Photoemission techniques

In this section we discuss two forms of photoemission spectroscopy, differentiated by the wavelength of light involved but sharing the goal of exploring surface electronic structure. In both cases, the strategy is to illuminate the sample with photons in the chosen energy range, and to analyse the electrons that are emitted in response.

Ultraviolet Photoemission Spectroscopy (UPS)

The central parameter in any photoemission experiment is the frequency of the illumination employed, which will typically be monochromatic to simplify the interpretation of results. Higher frequency photons are capable of exciting electrons from deeper in the valence band, or even from core states, while those just above the threshold frequency (with energy only just exceeding the work function) probe states just below the Fermi level. For studies of valence electrons in metals (and of the highest-energy occupied states of any molecules that may be adsorbed on the surface) it is often convenient to employ ultraviolet photons, with energies in the range 10–50 eV. Experiments are performed under ultra-high vacuum conditions, so that the photoemitted electrons travel freely from the surface to an electron energy analyser without being absorbed or scattered by intervening gas. This allows the intensity of emission to be recorded as a function of the photoelectron kinetic energy.

In a basic UPS experiment, no effort is made to analyse the angular distribution of photoelectrons—only their kinetic energy is of interest. To a first approximation, the number of photoelectrons recorded by the energy analyser at a particular kinetic energy, E_{kin}, should be proportional to the number of valence electrons (the density of occupied states) existing in the sample at a binding energy of E_b (measured relative to the Fermi level, as in Fig. 5.7) given by

$$E_b = h\nu - E_{kin} - \phi \tag{5.3}$$

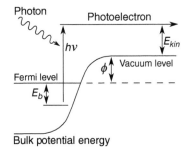

Fig. 5.7 Energies relevant to photoemission.

where $h\nu$ is the photon energy, and ϕ the work function. Note that a positive value of the binding energy corresponds to states below the Fermi level, and that binding energies less than zero cannot be measured by photoemission (because states above the Fermi level are unoccupied and hence cannot provide electrons to be emitted). Note also that this equation relates the photon energy, work function and binding energy on the assumption that no energy is lost by the photoelectron due to collisions on its way out of the material (elastic emission). If collisions occur, then the photoelectron may emerge at a lower kinetic energy than expected (inelastic emission) and some of the electrons that it has collided with may also be emitted from the solid (secondary emission). Fortunately, the distribution of inelastically emitted and secondary electrons is only a fairly weak function of kinetic energy, so this provides only a relatively featureless background spectrum upon which the more distinct peaks due to elastic photoelectrons are overlaid. This observation also underlines that it is attenuation of outgoing electrons that confers surface sensitivity upon photoemission experiments; incoming photons typically penetrate deep (perhaps microns) beneath the surface, but electrons excited at depth are unlikely ever to emerge (and certainly not elastically).

In an Angle-Resolved UPS (ARUPS) experiment, the energy analyser accepts photoelectrons from only a single direction, and the sample orientation is varied systematically to permit the intensity of emission to be recorded not only as a function of photoelectron kinetic energy, but also as a function of emission angle. Let us imagine that the analyser is arranged so as to detect electrons emitted at a polar angle of θ and some specific azimuthal angle (Fig. 5.8). Since the electrons arriving at the analyser are travelling through vacuum, their kinetic energy is simply given by the free-electron expression

$$E_{kin} = \frac{\hbar^2 k^2}{2m_e} \qquad [5.4]$$

where k is the photoelectron's wavenumber and m_e is its mass. It is thus straightforward to deduce the wavenumber from the measured kinetic energy. Armed with this value, we readily see that the experimental geometry implies

$$k_{xy} = k\sin\theta \qquad [5.5]$$

for the surface-parallel component of the wavevector (see Exercise 5.2).

Note that this expression for k_{xy} relates to the wavevector of the photoelectron as it travels through the vacuum *after* leaving the solid. It turns out, however, that electrons must conserve the surface-parallel component of their momentum (and therefore wavevector) during photoemission, because interaction with a perfectly smooth planar surface cannot impart momentum parallel to that surface,[9] and the incoming photon carries negligible momentum compared

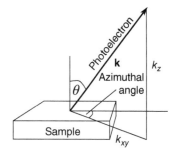

Fig. 5.8 Geometric parameters relevant to photoemission.

---

[9] For a classical analogy, one may imagine bouncing a ball elastically off a perfectly smooth planar wall: the component of momentum normal to the wall changes during the interaction, but the components of momentum parallel to the wall do not. Recall, however, that for quantum objects interacting with a crystalline surface, the wavevector is uniquely defined only to within the addition of an arbitrary surface reciprocal lattice vector, as noted in the discussion of LEED in Section 5.2.

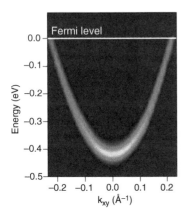

Fig. 5.9 ARUPS data showing, in greyscale proportional to the density of states, the Shockley surface state of Cu{111} (after Kim et al., 2012).

Shifts in the apparent binding energies of core electrons, as measured by XPS, may be caused by **initial-state effects** (relating to the different electrostatic potentials experienced by different atoms) or by **final-state effects** (relating to the different local relaxation of valence electrons in consequence of the photoemission process).

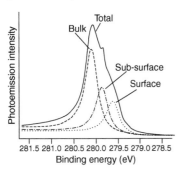

Fig. 5.10 XPS data from 3d core states at the Ru(10$\bar{1}$0) surface, showing core level shifts (Adapted from Baraldi et al., 2000).

with typical electron momenta. We therefore know that the photoelectron must have possessed a surface-parallel wavevector component equal to k_{xy} immediately prior to emission, but no such statement holds for the surface-normal component, k_z, which is not a conserved property. By repeating this analysis with intensities measured at different kinetic energies and polar emission angles, the angle-resolved experiment thus allows us to record the density of states associated with each particular value of k_{xy} along a specific azimuth. An example is shown in Fig. 5.9, highlighting the ability of ARUPS to elucidate surface electronic band structure (cf. Section 3.3).

X-ray Photoemission Spectroscopy (XPS)

In contrast to the valence electrons probed by UPS, the energies of core electrons show almost no dependence upon wavevector. This is because they are typically bound so tightly to their respective atomic nuclei that core states associated with one atom have virtually no overlap with the core states of neighbouring atoms. Solid materials thus feature a set of core states at very well-defined energies substantially below the Fermi level, accessible by illuminating the sample with X-rays. Since core states are typically bound tens or hundreds of eV below the Fermi level, X-rays of at least similar energy must be employed (per Eqn. 5.3).

One major use of XPS is simply to detect the presence of particular elements close to the surface. Each element has a characteristic set of core energy levels, which may be identified with peaks in the photoemission spectrum. More subtly, however, the technique can also yield information about the local chemical environment of the contributing atoms. For example, imagine a particular XPS peak deriving from atoms deep below the surface of an elemental solid. Sufficiently far below the surface, all such atoms occupy essentially identical local chemical environments, and so all contribute to this photoemission signal at a common binding energy. Atoms in the outermost layer, meanwhile, occupy positions with considerably different valence electron density than the bulk atoms, and so their contribution to this peak may be shifted both by an **initial-state effect** (due to the different local potential that they experience) and by a **final-state effect** (due to the different valence electron relaxation[10] that takes place upon photoemission). Exactly how big this surface core-level shift will be is hard to predict without performing sophisticated calculations, but displacements of 0.5 eV or more are certainly possible. In principle, one should also see a sequence of XPS peaks (not always well-resolved) associated with progressively deeper layers, gradually merging to form the main peak associated with the bulk atoms (Fig. 5.10).

More frequently, XPS is used to probe the nature of adsorbed species. Assuming no diffusion of adsorbates into sub-surface sites, the technique can

---

[10] The process of photoemission from a core state leaves behind a positively charged ion, which causes a rapid readjustment in the surrounding valence electron density. The energy change associated with this readjustment must be accounted for amongst the contributions that determine the final kinetic energy of the emitted photoelectron.

quantify relative coverage (since the peak intensity associated with a particular core state is proportional to the number of atoms possessing it). Moreover, sensitivity to the local chemical environment can indicate whether all atoms of a certain element occupy chemically equivalent positions or not. For example, if the adsorbate were a homonuclear diatomic, the presence of a single well-resolved peak associated with each core state would suggest a horizontal adsorption geometry, in which both atoms occupy broadly equivalent local chemical environments. In contrast, any peak-splitting would suggest a molecule binding upright on the surface, with the two atoms occupying rather distinct local chemical environments. In this way (and with empirical knowledge of the shifts typically associated with different local chemical environments) some quite complex chemistry may be mapped out (e.g. the distinction between oxygen atoms in hydroxyl, carbonyl, or carboxylate functional groups, and whether these are bound strongly to the surface or not).

Finally, it is worth noting a recent development that broadens the applicability of this technique quite dramatically. Attenuation of photoelectrons by collisions within gaseous media typically restricts the use of both UPS and XPS to ultra-high vacuum chambers, but clearly there are situations where one might reasonably ask whether the chemistry probed under such conditions really matches that taking place under ambient pressures in the real world. Innovative design of differentially pumped stages between the sample and the electron energy analyser, however, has recently permitted XPS to be performed under 'near-ambient' pressures (up to c.30 mbar) and this new capability promises to have major impact upon studies aimed towards fields such as corrosion and catalysis, where pressure is likely to be of more than passing importance to the surface chemistry. The interested reader is referred to Salmeron and Schlögl (2008) and Arble et al. (2018) for comprehensive reviews.

5.5 Auger Electron Spectroscopy (AES)

In Auger electron spectroscopy, the sample is struck by a beam of electrons having sufficient energy (typically 2–10 keV) to dislodge a core electron (the **primary electron**) from an atom by direct collision, creating a hole in its electronic structure at energy E_1 (see Fig. 5.11a). Such a **core hole** is short-lived, though, because an electron from a higher energy state (energy E_2) can readily drop into it. Nevertheless, the delay between creation and re-filling of the hole is typically long enough that they can be considered as two separate events, which must both individually conserve energy. The former process involves a variable (and unknown) degree of energy transfer between the incident electron and the core electron,[11] which makes quantitative analysis impossible. The latter process, however, is highly diagnostic of the atom's elemental identity.

[11] It ought to be stressed that the Auger process (arguably the Meitner-Auger process) may be instigated by any event capable of ionising a low-lying core state; e.g. the absorption of a photon during an XPS experiment. As a rule, however, the spectroscopic technique known as AES implies that a high-energy electron beam has been used to stimulate Auger emission.

AES is instigated by the emission of a **primary electron**, giving rise to a **core hole** that will be filled by the so-called **down electron**. The upshot is emission of an **Auger electron**, whose energy will be diagnostic of the element involved.

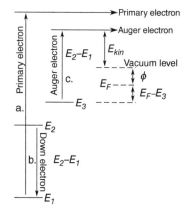

Fig. 5.11 Energies and processes relevant to Auger spectroscopy.

Consider the refilling of a recently created core hole by an electron from a specific higher energy core state (the **down electron**) implying a drop in the potential energy of the system by a precisely defined amount (equal to the energy difference, $E_2 - E_1$, between the two core states involved, see Fig. 5.11b). Clearly this drop in potential energy must be accompanied by a corresponding increase in the energy of some other part of the system, in order to conserve total energy, and one way in which this can occur is for another core electron (energy E_3) to be ejected from the atom (see Fig. 5.11c). If this electron (the **Auger electron**) escapes from the material altogether, its ultimate kinetic energy (E_{kin}) will necessarily equal the energy difference between the first two core states ($E_2 - E_1$) minus the sum of the sample work function (ϕ) and the binding energy of its original core state ($E_F - E_3$). That is,

$$E_{kin} = (E_2 - E_1) - (E_F - E_3) - \phi$$
$$= (E_2 + E_3 - E_1) - (E_F + \phi)$$

[5.6]

assuming no energy is lost through collisions during the escape.[12]

Since the energy $E_2 + E_3 - E_1$ depends only upon the relative energies of core states, which are only mildly affected by the local chemical environment of the atom involved, each element gives rise to a unique and diagnostic Auger spectrum. Different values of E_F merely produce a fairly small material-specific offset. The peaks in this spectrum are generally rather sharp, but they are of relatively low intensity and overlie a sloping background due to secondary electron emission. To help distinguish these features, it is common practice to plot not the number of electrons emitted with a given kinetic energy, but rather the energy derivative of this quantity. A relatively featureless sloping background contributes only a small and fairly constant value to this derivative, whereas even a weak peak contributes a significant positive excursion to the derivative on the low-energy side (corresponding to the up-slope of the peak in emission) and a similarly significant negative excursion to the derivative on the high-energy side (corresponding to the down-slope of the peak in emission). The intensity of Auger emission is generally measured as the so-called **peak-to-peak height** of the derivative feature; that is, the difference in the differentiated signal between the top of the positive excursion and the bottom of the negative excursion.

Some typical Auger spectra are shown in Fig. 5.12, in which one can readily identify several strong derivative features corresponding to emission from substrate atoms, together with a number of additional derivative features due to emission from impurities. In quantitative work, it is common to report the intensity ratio of adsorbate- or impurity-related features against substrate-related features, in order to suppress fluctuations that may arise solely from variations in the incident electron flux. It is therefore possible not only to identify but also to quantify the impurities that may be present in a given sample.

AES spectra are often presented by plotting the first derivative of the signal with respect to kinetic energy. Intensity of Auger emission is then gauged from the **peak-to-peak height** of sharp features in the derivative spectrum.

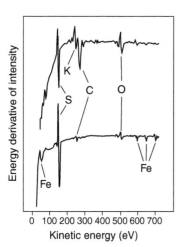

Fig. 5.12 Typical Auger spectra (from an FeS$_2${100} surface with residual impurities). Top and bottom traces are before and after cleaning via sputtering [after Temprano et al, 2017]

---

[12] It should be remembered that the relevant core energies refer to an atom in which one of the lower-lying core electrons has already been removed, rather than the core energies of the unperturbed atom.

5.6 Near-Edge X-ray Absorption Fine Structure (NEXAFS)

We have already seen that X-rays may be absorbed in processes involving photoemission from core states. By measuring the kinetic energy of such photoelectrons, one may reconstruct corresponding binding energies, and hence not only determine which elements are present, but also infer aspects of their local chemical environment (Section 5.4). Can we, however, gain additional information by studying the absorption coefficient of the X-rays themselves, and if so how might we measure this?

Rather than monitoring the absorption of X-rays directly, it is usual to measure instead the so-called **total electron yield**. The sample is connected to Earth via an ammeter, and the drain current necessary to maintain electrical neutrality is recorded during illumination by the X-ray source. In this way, one obtains the total rate of electron emission. Importantly, this includes not only those electrons emitted via the elastic process measured in XPS, but also electrons emitted through Auger-type processes, not to mention inelastically scattered and secondary electrons. The total electron yield is thus approximately proportional to the overall X-ray absorption coefficient, even counting occasions where the primary photoelectron does not ultimately exit the solid but instead enters an unoccupied state between the Fermi level and the vacuum level.

> In collecting NEXAFS data, the **total electron yield** (measured by means of the drain current) is typically used as a convenient proxy for the X-ray absorption coefficient.

If the X-ray photon energy is progressively increased, one observes distinct upward jumps in the absorption coefficient whenever the binding energy of a new core state is exceeded. Below such a threshold, incoming photons have insufficient energy to excite electrons from this particular core state into an unoccupied state, but above it excitation becomes possible. We thus anticipate a series of 'edges' in the absorption coefficient, and it is the fine structure that arises close to these edges that we analyse in the NEXAFS technique.[13]

Let us illustrate by considering one particular absorption edge, corresponding to excitation from a specific core state. The change in X-ray absorption coefficient at the edge itself will be proportional to the density of states just above the Fermi level (the number of available unoccupied states into which the core electron may be excited) and as the photon energy is increased so the absorption coefficient will broadly mimic the variation in density of states at correspondingly higher energies above the Fermi level.[14] In addition, however, the absorption coefficient will also be proportional to the integral

$$\int \psi_c^*(\mathbf{r})\mathbf{r}.\hat{\mathbf{e}}\,\psi_u(\mathbf{r})\mathrm{d}\mathbf{r} \qquad\qquad [5.7]$$

---

[13] Physicists make an anagram out of NEXAFS—omitting the word 'fine'—and refer to the same technique as XANES. One should not, however, confuse either NEXAFS or XANES with Extended X-ray Absorption Fine Structure (EXAFS) or Surface EXAFS (SEXAFS), both of which make use of features beyond the near-edge region of the spectrum, and neither of which are included in this book.

[14] It is, of course, almost as important to study the unoccupied states of a system as it is to study the occupied states. One would not dream, for example, of considering only the HOMO when discussing molecular reactivity; the LUMO is also essential to a satisfactory understanding. The same is equally true of electronic structure at surfaces.

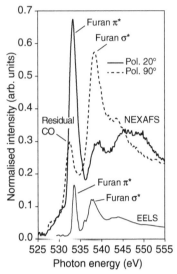

Fig. 5.13 O 1s NEXAFS spectra of furan adsorbed on Pd{111}, with polarisation angles quoted relative to the surface normal (after Knight et al., 2008). An electronic density of (unoccupied) states, measured by Electron Energy Loss Spectroscopy (EELS), is shown for comparison (after Duflot et al., 2003).

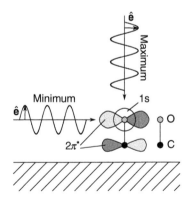

Fig. 5.14 Polarisation dependence in NEXAFS from the O 1s core state.

where $\psi_c(\mathbf{r})$ and $\psi_u(\mathbf{r})$ are the eigenfunctions of the core and unoccupied states respectively, and where $\hat{\mathbf{e}}$ is the unit polarisation vector of the X-ray (defining the direction in which the electric field vector oscillates). Accordingly, absorption will be significant only if there is considerable spatial overlap between the unoccupied eigenfunction and the core eigenfunction, so NEXAFS effectively probes only the *local* density of states in the vicinity of the atom whose core state is excited. For example, in studying an adsorbate such as furan (C_4H_4O) one can address separately the fine structure close to the carbon and oxygen 1s edges, gaining information in one case about the density of states local to the carbon atoms, and in the other about the density of states local to the oxygen atom (Fig. 5.13). NEXAFS is thus not only highly surface sensitive, but is even specific to different parts of a single molecule.

In fact, Eqn. 5.7 suggests yet a further refinement to the analysis of NEXAFS data. Let us simplify (without loss of generality) by taking the origin of our co-ordinate system to lie at the nucleus of the atom whose edge we happen to be studying. If we also restrict ourselves to core states of spherical symmetry, then the eigenfunction $\psi_c(\mathbf{r})$ will always be a symmetric function across any plane passing through the origin. The product $\mathbf{r}.\hat{\mathbf{e}}$, on the other hand, will necessarily be an antisymmetric function across a plane perpendicular to the polarisation vector and passing through the origin. It follows, therefore, that the integrand of Eqn. 5.7 will be a similarly antisymmetric function if $\psi_u(\mathbf{r})$ (or at least that part of it lying within the region for which the core state wavefunction is non-negligible) is symmetric across that same plane, in which case the absorption coefficient must go strictly to zero. By systematically varying the polarisation vector of the incoming X-rays, therefore, it is possible to probe the symmetry of those unoccupied orbitals that contribute to peaks in the NEXAFS spectrum (see Exercise 5.3). A relatively simple example may be found in the case of CO adsorption. As shown in Fig. 5.14, the symmetry of the $2\pi^*$ orbital means that its contribution to the O 1s edge is negligible when the X-ray polarisation vector, $\hat{\mathbf{e}}$, is perpendicular to the surface, and maximised when it is parallel to the surface, assuming an upright adsorption geometry. Conversely, if the adsorption geometry of the molecule is *not* known, but its electronic structure *is*, then NEXAFS can be a powerful tool to explore molecular orientation with respect to the surface.

5.7 Molecular beam techniques

In most experiments where atoms or molecules are deposited upon a solid surface, this is achieved simply by exposing the sample to gas of an appropriate composition. In terms of the kinetics of adsorption, the particles within the gas strike the surface with a range of velocities (and hence translational energies) dictated by the Maxwell–Boltzmann distribution, and at a rate linked to the pressure at the surface. Two drawbacks are immediately evident: firstly, that one cannot achieve an impingement rate comparable with that found at ambient pressures without compromising the ultra-high vacuum conditions necessary for most surface science techniques; and secondly, that it is impossible to determine whether it is the low-, medium-, or high-energy particles that are responsible for whatever phenomena are observed.

The first issue may be addressed by employing an **effusive molecular beam**, whereby gas is introduced into the experimental chamber not simply by opening a valve, but instead by permitting it to pass through a narrow orifice or nozzle. If the nozzle is sufficiently small, then the gas may be thought of as entering the chamber in a directed beam, and this directionality can be further enhanced through collimation. The beam thus creates a region close to the surface within which the local pressure is substantially higher (perhaps by several orders of magnitude) than pertains throughout the rest of the chamber. It should be stressed, however, that particles within an effusive beam continue to be governed by the kinetic theory of gases, so their velocities remain thermally distributed.

In order to address the second issue, it is typically necessary to make use of a narrower nozzle and a higher upstream pressure, in which case it is possible to generate a **supersonic molecular beam** exhibiting collision-free laminar flow in the central region of its expansion (Fig. 5.15). With pressures of P_n and P_0 on the upstream and downstream sides of the nozzle, conditions for supersonic behaviour (in which flow velocity exceeds the local speed of sound) will be maintained over a distance approximated by $\sqrt{4P_n/9P_0}$ nozzle diameters. Within this supersonic region, which may be isolated by means of a skimmer, the constituent particles do not follow the Maxwell–Boltzmann distribution, but instead each species is practically monoenergetic. The translational energy for species i is then dictated solely by the nozzle temperature, T_n, according to the equation

$$E_i = \left(\frac{M_i}{M}\right)\frac{C_p R T_n}{(C_p - C_v)} \tag{5.8}$$

where C_p and C_v are the constant-pressure and constant-volume molar heat capacities of the gas, M_i is the mass of species i, and M is the mean mass of the particles in the beam. By systematically varying the nozzle temperature, therefore, it is possible to investigate the role that translational energy may have upon surface adsorption and reaction kinetics (Sections 4.2 and 4.5).

An **effusive molecular beam** is a collimated stream of gas that provides a directed and intense source of molecules within a vacuum chamber, but its constituent particles conform closely to predictions from the kinetic theory of gases. In a **supersonic molecular beam**, by way of contrast, the translational motion of particles is nearly monoenergetic, while rotational and vibrational motion is strongly quenched.

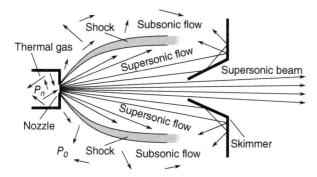

Fig. 5.15 Generation of a supersonic molecular beam.

One complicating factor in such studies, however, relates to the vibrational and rotational motion of the molecules within the beam. In general, the constituent particles of a supersonic beam exhibit thermal distributions of vibrational and rotational energy, but with effective temperatures for these degrees of freedom considerably lower than the nozzle temperature. That is, they may be said to be vibrationally and rotationally *cold*. Nevertheless, an increase in nozzle temperature will certainly increase not only the translational energy of the beam, but also its vibrational and rotational energies too. If one wishes to vary *only* the translational energy, it is often more appropriate to 'seed' the beam with a noble gas, thus varying the mean particle mass that appears in Eqn. 5.8 without affecting the chemistry of the beam; seeding with a noble gas that lowers the mean mass of the beam leads to an increase in the translational energy of the other beam species, while seeding with a noble gas that raises the mean mass would have the opposite effect.

It is, of course, possible to precisely balance a change in nozzle temperature against a contrary change in the seeding of a supersonic beam, in order to maintain a constant translational energy whilst varying the vibrational and rotational energies. This can be a useful way of gauging the effect of non-translational energy in surface adsorption and reaction. Recently, however, some groups have developed sophisticated methods of directly exciting or selecting specific types of vibrational or rotational motion, through the use of laser excitation or time-of-flight analysis.

The King and Wells technique

One of the most fundamental applications of supersonic molecular beams in surface science is the measurement of sticking probabilities, most often carried out by some variation of a technique first described by King and Wells (1972). In this approach, the path of the beam from nozzle to sample is intercepted by an inert 'flag' that may be withdrawn and reinserted at will. The operation of this flag, and the interpretation of the results obtained when using it, are best explained with reference to a hypothetical example, shown in Fig. 5.16.

Initially, the beam is inoperative and the chamber is pumped down to its base pressure (P_A). The experiment proper begins when the beam is switched on, but at first it strikes only the interposed inert flag and not the sample; all of the molecules within the beam are therefore scattered into the chamber, and the pressure measurably rises (P_B). Once the pressure has stabilised (when the pumping rate equals the beam flux), the flag is abruptly withdrawn out of the beam path, so that molecules now strike the sample. Since some fraction of these molecules stick to the sample, rather than being scattered, the pressure in the chamber drops (P_C). Conveniently, the ratio $(P_B - P_C)/(P_B - P_A)$ gives a direct measure of the fraction of molecules within the beam that stick to the surface, which is to say that it equates to the initial value (the clean-surface value) of the sticking probability, s, defined in Section 4.2.

As adsorption proceeds, the surface coverage gradually increases and the sticking probability consequently varies, but its value may be obtained from

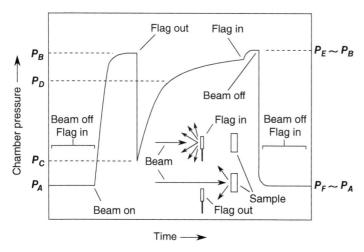

Fig. 5.16 Schematic of a King and Wells experiment and example pressure trace.

the pressure observed at any arbitrary moment (P_D) by evaluating the ratio $(P_B - P_D)/(P_B - P_A)$. At the end of the experiment, the flag is reinserted into the beam path, and the chamber pressure rises to a value (P_E) that should ideally be equal to that observed immediately prior to its previous withdrawal. Finally, the beam is switched off and the chamber returns to its base pressure (P_F). If the molecular flux within the beam is known, and if it may reasonably be assumed that adsorbates have insufficient time to diffuse out of the region of the surface struck by the beam, then knowledge of the changing sticking probability through the course of the experiment permits one to calculate the absolute coverage within the beam's footprint at any moment in time. Doing so, it becomes possible to construct a plot showing the variation of sticking probability as a function of surface coverage, as exemplified in Fig. 5.17 (see Exercise 5.4).

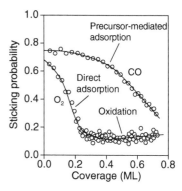

Fig. 5.17 Room-temperature sticking probability measurements for CO and O_2 on Co{110} (after Liao et al., 2012–13).

Single-Crystal Adsorption Calorimetry (SCAC)

It is worth examining briefly a rather specialised method that builds upon the molecular beam techniques discussed above and permits direct measurement of adsorption heat. In a SCAC experiment, a supersonic beam is operated in pulsed mode, gradually building up coverage of a particular molecular species on an exceptionally thin (c.100 nm) single-crystal sample. A flux gauge is used to estimate the number of molecules per pulse, and the sticking probability is measured (by a variant of the King and Wells method) so that the number of molecules actually deposited per pulse may be accurately determined. The very low heat capacity of the sample ensures that it experiences a small but measurable increase in temperature upon the arrival of each pulse, and this too may be measured by some convenient method (e.g. with an infrared bolometer or pyroelectric strip in proximity or contact with the back of the sample). Before and after each experiment, a pulsed laser of known power is used for calibration, effectively establishing the heat capacity of the sample, so that the measured temperature change upon

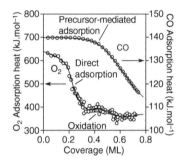

Fig. 5.18 Room-temperature adsorption heat measurements for CO and O_2 on Co{110} (after Liao et al., 2012–13).

adsorption can be converted to a corresponding quantity of absorbed heat. Collating all of this information permits calculation of the heat released per molecule upon adsorption, as a function of surface coverage. Some typical results are shown in Fig. 5.18 (see Exercise 5.4). The advantage of this approach over isosteric measurements (Section 1.8) is that it works for irreversible as well as for reversible adsorption, while in comparison with kinetic estimates (Section 4.3) it works for activated as well as for non-activated adsorption.

5.8 Vibrational spectroscopies

Identification of chemical species via their vibrational spectra is a staple of gas-phase analysis, resting upon the absorption of infrared radiation by functional groups with well-known fundamental frequencies. The intensity of a given vibrational normal mode is proportional to the square of the dynamic dipole moment, defined as the derivative of the molecule's dipole moment with respect to displacement along the normal-mode coordinate.[15] Infrared spectroscopy is thus a rather powerful technique, capable not only of distinguishing between chemical species, but also of quantifying their concentrations, and it is hardly surprising that surface chemists would wish to make use of it. Here we review some of the more successful approaches that have been taken.

Reflection Absorption InfraRed Spectroscopy (RAIRS)

In RAIRS experiments on metals, the surface is illuminated at near-grazing incidence (typically at least 80° off-normal) by infrared radiation, and reflected light is detected at the specular angle. A spectrometer, placed between source and sample, permits spectral analysis with a typical resolution of a few wavenumbers.[16] When working with a sample under ultra-high vacuum conditions (preferable for high-sensitivity experiments), the source, spectrometer, and detector nevertheless usually sit outside of the vacuum environment, so it is vital that they be purged with infrared inactive gas (e.g. dry nitrogen) to ensure that absorption within these sections of the beam path is minimised. Inside the vacuum chamber, the only location at which absorption should be expected is, of course, the surface itself.

The use of grazing incidence in RAIRS is motivated by two considerations. Firstly, such a geometry maximises the surface area (and hence the number of adsorbates) probed by a beam of fixed diameter, thus maximising the surface signal in relation to that arising from residual species in the space above the sample.

---

[15] In highly symmetric molecules, some normal modes may be invisible to infrared spectroscopy, due to a vanishing dynamic dipole moment—a statement of the infrared selection rule.

[16] The most common choice of spectrometer in RAIRS applications is the Fourier Transform InfraRed (FTIR) type, in which the resolution is inversely related to the distance travelled by an internal moving mirror. The resolution can often be selected by the user, but since the speed of the moving mirror is typically limited by practical constraints, the inevitable corollary of improved resolution is an increase in acquisition time for each spectrum.

Secondly, and more subtly, grazing incidence ensures that one of the polarisation components of the beam lies close to the surface normal.

When light reflects from a metal surface, that component of the electric field lying perpendicular to the surface is doubled in strength within the near-field region (i.e. within about one wavelength from the surface) while that component of the electric field lying parallel to the surface is extinguished (Fig. 5.19). From this, it follows that only the polarisation in which the electric field oscillates within the sagittal plane (the plane containing both the wavevector and the surface-normal vector) can be absorbed by surface vibrational modes, and hence that the orthogonal polarisation is entirely redundant for our purposes. Moreover, it further follows that only the surface-normal component of a vibrational mode's dynamic dipole moment couples to the impinging radiation. This, in turn, implies the existence of a **surface selection rule** to the effect that any vibrational mode lacking a surface-normal component to its dynamic dipole moment will be invisible to RAIRS (see Fig. 5.20). This rule can be viewed as a problem, in the sense that certain diagnostic vibrational modes known from gas-phase species may be absent from surface absorption spectra, but it can equally be considered fortuitous, in that the absence of a particular vibrational mode that otherwise ought to be present may provide a useful clue as to the orientation of an adsorbed molecule. It is important to note, however, that even vibrations taking place wholly within the plane of the surface can give rise to surface-normal dynamic dipole moments—the motion of charge, not atoms, is key (see Exercise 5.5).

It ought to be mentioned here that RAIRS experiments are typically limited by the availability of suitable sources and/or detectors of infrared light, and that 600 cm^{-1} may consequently be considered an optimistic lower bound on the frequencies for which it provides reliable results. This allows access to most intramolecular vibrational frequencies of interest, but does mean that frustrated translations and rotations cannot be studied by this technique. For access to these, and to phonon modes of the surface itself, one might turn to High-Resolution Electron Energy Loss Spectroscopy (HREELS) or helium scattering techniques, omitted from this book on grounds of brevity. The other major drawback in RAIRS, however, relates to the preference for ultra-high vacuum conditions, to avoid undue absorption of the infrared beam on its path to and from the surface (and consequent signal-to-noise issues). Two approaches to circumvent this limitation are outlined briefly below.

Attenuated Total Reflectance (ATR)

In an ATR experiment, infrared light is passed through an infrared-transparent prism of high refractive index, in such a manner as to undergo total internal reflection (often several times) before leaving the prism and impinging upon a suitable detector. Whilst the electromagnetic field within the prism behaves according to our usual expectations of a propagating wave, it must also be understood that there will exist a non-propagating evanescent wave outside of the prism, oscillating at the same frequency as the light driving it, but decaying

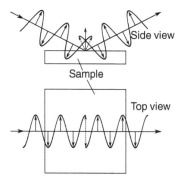

Fig. 5.19 Side and top views of reflection from a metal surface. The surface-parallel electric field components in both cases experience a 180° phase shift, while the surface-normal component does not.

The **surface selection rule** limits detection of vibrational modes by RAIRS to those that are associated with a dynamic dipole moment perpendicular to the surface.

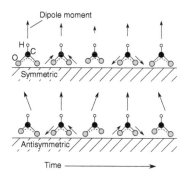

Fig. 5.20 Symmetric and antisymmetric vibrations of adsorbed formate. Oscillation of the surface-normal dipole component occurs only in the symmetric case.

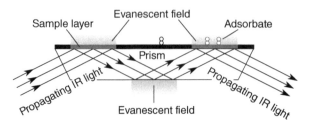

Fig. 5.21 Geometry for an ATR experiment.

exponentially with distance from each reflecting surface (see Fig. 5.21). If we now coat one reflecting surface with a sufficiently thin layer of our sample material, it may be possible that the evanescent wave, although undoubtedly screened, nevertheless retains some appreciable amplitude at (and beyond) the free surface of the coating. In such circumstances, coupling of the evanescent wave with vibrations of adsorbates at the free surface will result in a reduction in amplitude of the reflected beam at the corresponding frequency. In effect, the arrangement functions as a RAIRS experiment carried out from beneath the surface, and the apparatus and methods of analysis employed will look much the same. Since the propagating wave never enters the space above the coated layer, it matters not at all whether that space is held at vacuum, occupied by gas, or indeed filled with liquid. In this way, the ATR technique offers access to a range of surface conditions not compatible with standard RAIRS experiments. Its most serious limitation, however, is that it can only be used in cases where the surface we wish to study can be formed by deposition onto the prism material. Studies of single-crystal samples, for example, are not generally possible via this technique.

Sum-Frequency Generation (SFG)

The phenomenon of SFG arises due to non-linear effects in the passage of light through polarisable media. A mathematical treatment is beyond the scope of this book, but the basic idea may be quite readily summarised. As a beam of light travels through a material, it necessarily creates a polarisation field in its vicinity, which oscillates not only at the frequency of the light but also at higher harmonics (i.e. multiples of the fundamental frequency). These higher harmonics are usually ignored in elementary treatments, but they in fact imply that passage of light through a polarisable medium necessarily generates subsidiary light at the harmonic frequencies. If a second beam of light, perhaps with a different frequency to the first, also passes through the same polarisable material, the combined polarisation due to both beams will, in general, now oscillate not only at the two fundamental frequencies, but also at linear combinations of the fundamental frequencies. To first-order, only the fundamental frequencies are found, but to second-order the fundamentals are accompanied both by the second harmonics of the two fundamental frequencies, and by oscillations at the sum and at the difference of the two fundamental frequencies. The important exception to this rule, however, is that all of these non-linear effects vanish when the medium exhibits inversion symmetry, and this turns out to be key to their practical utility.

A typical SFG experiment involves impinging two beams of laser light upon the surface of interest (see Fig. 5.22). One of these beams must be monochromatic, with a frequency usually chosen to lie within the visible region of the electromagnetic spectrum. The other beam will lie within the infrared region, so that it may couple to vibrational motion, and should be tunable, to permit spectroscopic measurements. Laser light (often pulsed) is necessary in order to achieve sufficient intensity at the surface for substantial second-order effects to occur. A suitable detector is then arranged to monitor light emitted from the surface at frequencies consistent with the sum of those belonging to the visible and infrared lasers. Crucially, the space above the surface, even if occupied by gas or liquid, necessarily exhibits inversion symmetry, meaning that no sum-frequency signal can arise from this region. Similarly, we can usually rule out a sum-frequency signal arising from the bulk material beneath the surface[17] and can, in that case, be certain that the SFG signal arises solely from the near-surface region, where inversion symmetry is necessarily broken. Analysis of the SFG signal, perhaps to obtain information about vibrations of adsorbates at the surface, is inevitably more complicated than for RAIRS, since it involves consideration of the polarisability tensor, but nevertheless the identity, conformation, and orientation of functional groups may reliably be determined from the data obtained.

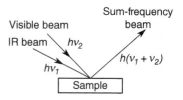

Fig. 5.22 Geometry for an SFG experiment.

5.9 Density Functional Theory (DFT)

Since the last years of the twentieth century, theoretical studies of surfaces have been dominated by DFT. In contrast to traditional computational chemistry, in which electronic wavefunctions are central, DFT's success lies in focussing instead upon the system's ground-state electron density. According to two foundational theorems of Hohenberg and Kohn (1964) it may be asserted (i) that the many-electron ground-state wavefunction of a system is a unique functional of its electron density, and (ii) that the ground-state electron density is that which minimises the total energy functional of the system.[18] Once the ground-state electron density is known, any other ground-state properties may, in principle, be computed with relative ease. All that is missing from this happy state of affairs is a practically applicable and fully accurate expression for the total energy functional. Fortunately, the efforts of Kohn and Sham (1965) rapidly supplied a work-around for this problem, which we shall now review.

---

[17] The exception would be if we happen to be dealing with one of those crystalline materials without inversion symmetry.

[18] In mathematics, a functional is a function of a function. Taking the present context as our example, the electron density is a function of the position variable (i.e. each position in space is associated with a specific value of the electron density) and both the ground-state many-electron wavefunction and the total energy of the system are functionals of the electron density (i.e. they each depend upon all of the electron density values throughout the space occupied by the system).

The Kohn–Sham formulation

The cornerstone of practical DFT implementation is the Kohn–Sham equation, derived from the Hohenberg–Kohn theorems and stated as

$$\left(-\frac{\hbar^2}{2m_e}\nabla^2 + V_{KS}(\mathbf{r})\right)\varphi_n(\mathbf{r}) = \varepsilon_n\varphi_n(\mathbf{r}) \qquad [5.9]$$

with $V_{KS}(\mathbf{r})$ representing the Kohn–Sham potential. The form of Eqn. 5.9 clearly mimics the Schrödinger equation, but the eigensolutions ε_n and $\varphi_n(\mathbf{r})$ ought not to be interpreted uncritically as having the physical meaning of electronic wavefunctions.[19] In practice, they often do approximate the eigensolutions obtained by non-DFT methods (and are almost invariably treated as such) but strictly they appear here merely as a side-effect of the mathematical derivation carried through by Kohn and Sham. Nevertheless, the electron density of the system may be generated by summing the square moduli of those $\varphi_n(\mathbf{r})$ whose corresponding ε_n fall below some value that we might reasonably interpret as dividing occupied from unoccupied electronic states, consistent with the total charge of the system.

The first term in the Kohn–Sham Hamiltonian represents the kinetic energy operator for a system of non-interacting electrons with eigenfunctions $\varphi_n(\mathbf{r})$ and mass m_e, while the second term represents an effective potential within which they move. The latter may be broken down as

$$V_{KS}(\mathbf{r}) = V_{ext}(\mathbf{r}) + V_H(\mathbf{r}) + V_{xc}(\mathbf{r}) \qquad [5.10]$$

where $V_{ext}(\mathbf{r})$ represents any externally applied potential, $V_H(\mathbf{r})$ represents the so-called Hartree potential, and $V_{xc}(\mathbf{r})$ represents the exchange-correlation potential (the derivative of a corresponding exchange-correlation functional with respect to electron density). The external potential typically includes any interactions between electrons and the nuclei within the system, along with the effect of any applied electric field. The Hartree potential includes Coulombic interactions between each electron and the prevailing electron density distribution of the system. The exchange-correlation potential, meanwhile, accounts for two important effects: firstly, it incorporates the exchange interaction, which has its origin in the exclusion principle; and secondly, it includes the correlation interaction, which arises because the dynamic behaviour of electrons is not fully captured by the non-interacting kinetic term and the other potentials. If the correct form of the exchange-correlation potential were known, solution of the Kohn–Sham equation would be possible, yielding knowledge of the ground-state electron density, and hence knowledge of any ground-state property of the system we might desire to calculate. Unfortunately, the correct form of the exchange-correlation potential is not known, except in the case of a perfectly uniform electron gas, so we must invariably employ some guile.

---

[19] We use φ in place of ψ throughout this section, as a visual reminder that Kohn–Sham eigenfunctions are a little different from the electronic eigenfunctions we normally encounter.

The least sophisticated approach to a non-uniform system is the so-called **Local Density Approximation (LDA)**, which embodies the simplifying assertion that each point in space contributes to the overall exchange-correlation functional an energy consistent with an electron gas of *uniform* density equal to the *local* electron density of the actual system at that point. Although this approximation works surprisingly well in many cases, it is often insufficient in systems where the electron density varies sharply across space, and since the 1980s a variety of **Generalised Gradient Approximations (GGA)** have been developed. These GGA functionals share the characteristic that they depend not only upon the local electron density at each point, but also upon its gradient; they differ in the details of how the gradient-dependence is parametrised, and in what kind of constraints are employed in fixing the values of the associated parameters.[20]

Amongst the properties obtainable via DFT we may include the forces that act upon atomic nuclei. It is therefore possible to make an educated guess at the structure of a molecule or solid, calculate any unresolved forces acting upon its nuclei, move those nuclei in response, and then iterate the procedure until the forces fall below some user-specified threshold. The result of such a geometry optimisation process is a structure representing at least a local energy minimum of the system, but it cannot be definitively identified as a global energy minimum. We are justified in surmising that it *may* be the global minimum only after repeating the calculation with a sufficient number of different starting geometries and thus gaining some insight into the overall energy landscape. Once local energy minima have been identified, the ability to calculate forces acting when the corresponding structures are perturbed permits calculation not only of vibrational spectra (frequencies and normal modes) but also of transition states (and hence activation energies). Molecular dynamics simulation based upon first-principles DFT provides another use for calculated forces, although computational expense makes this a relatively rare approach (compared with force-field methods) in the context of surface studies.

Application to surfaces

Most DFT calculations for surface systems (though by no means all) make use of periodic boundary conditions, permitting the Kohn–Sham wavefunctions to be conveniently expanded in a set of plane waves consistent with Bloch's theorem (cf. Section 3.3). That is, we have

$$\varphi_{nk}(\mathbf{r}) = \sum_{\mathbf{G}} A_{nk}^{\mathbf{G}} \exp(i(\mathbf{k} + \mathbf{G}).\mathbf{r}) \qquad [5.11]$$

where $\mathbf{k}$ is a wavevector within the 1BZ, and where the $A_{nk}^{\mathbf{G}}$ coefficients must be determined such that the Kohn–Sham equation (Eqn. 5.9) is satisfied. The quality of this plane-wave expansion is controlled by the extent of the sum over reciprocal lattice vectors, $\mathbf{G}$, which ought, in principle, to encompass the

The simplest description of the exchange-correlation functional in a non-uniform system is that provided by the **Local Density Approximation (LDA)** but this works less well when electron densities vary sharply. For molecules and surfaces, it is usually better to make use of **Generalised Gradient Approximations (GGA)** of which several different varieties are readily available.

[20] We omit consideration of so-called hybrid or meta functionals, which have not, as yet, found widespread popularity within the surface science community.

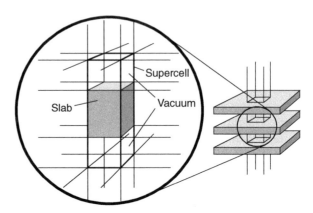

Fig. 5.23 Slab/supercell geometry for DFT with periodic boundary conditions.

entire (infinite) reciprocal lattice. In practice, however, the sum is truncated such that the only plane waves included are those for which the kinetic energy $\hbar^2(\mathbf{k}+\mathbf{G})^2/2m_e$ falls below some specified cutoff value. Since large wavenumbers correspond to short wavelengths, the imposition of this cutoff amounts to a limit on the amount of fine detail that can be captured by the expansion, but use of pseudopotentials[21] can dramatically reduce the amount of fine detail necessary for an adequate description of the valence states. Suitable cutoffs are largely dictated by the particular elements present in the system studied, or rather by their chosen pseudopotentials.

Use of periodic boundary conditions necessitates, of course, a model of the surface system that is itself periodic. In the two dimensions parallel to the surface this presents little difficulty, at least in the case of a crystalline material; one simply models a patch of surface commensurate with the two-dimensional unit cell of the clean, reconstructed, or adsorbate-covered surface, as appropriate. In the dimension perpendicular to the surface, however, one must invoke an artificial periodicity, modelling a slab of material with vacuum on either side. The slab thickness must be sufficient that opposite sides interact negligibly with one another, and the vacuum thickness must similarly be sufficient that one side of the slab interacts negligibly with the other side of its neighbouring periodic image (see Fig. 5.23).

In some circumstances—for example, when calculating work functions or surface stress—it is most convenient to permit atoms on both sides of the slab to relax in response to the calculated forces, so that the relaxed slab effectively models two symmetrically equivalent surfaces, one on either side; it is usual, nonetheless, to constrain some number of atomic layers within the core of the slab to their ideal bulk positions, preventing an unphysical inward collapse. In other circumstances, especially when adsorbates are to be included, it is

---

[21] Here we note only that one may replace the true interaction potential between a valence electron and an ionic core with a carefully constructed and much weaker potential—the pseudopotential—that nevertheless reproduces the correct electronic behaviour beyond some specified distance from the ion.

common to constrain the entirety of the slab apart from a few layers on just one side; this then becomes the surface of interest, to which adsorbates may be attached or reconstructions applied if desired, while the frozen side of the slab is now an explicitly unphysical surface whose properties should be ignored. Whichever approach is taken, the thickness of the slab and the vacuum gap between slabs ought to be tested for convergence. As a rule of thumb, 10–20 Å of vacuum will usually prove sufficient, unless particularly large molecules are to be adsorbed, while the necessary slab thickness will be material-dependant but often in a similar range.

Once the full three-dimensional periodicity is settled upon, the calculation can proceed within the corresponding unit cell, usually referred to as a supercell in view of its size relative to the bulk unit cell. The chosen supercell then implies a corresponding reciprocal lattice, and the nature of this will influence the cost of the DFT calculation. Specifically, the larger the supercell the more closely spaced will be the reciprocal lattice vectors, and hence the more plane waves will fall within the truncation limit implied by a given kinetic energy cutoff. Making the supercell larger—even when this involves only increasing the vacuum region without adding any more atoms to the system—will substantially increase the cost of the calculation. At the same time, it ought to be understood that evaluation of the ground-state electron density will require summation over the square moduli of Kohn–Sham eigenfunctions, $\varphi_n(\mathbf{r})$, drawn from throughout the 1BZ. In practice, one samples the 1BZ at a small number of $\mathbf{k}$-points, carefully chosen to be as representative of the full electronic structure as possible. The length of the supercell in its surface-normal dimension implies a very flat 1BZ in the corresponding dimension, meaning that only a single layer of points is needed, but the number of points within that layer necessary to adequately sample the surface-parallel dimensions will depend upon both the lateral size of the supercell (larger size needs fewer $\mathbf{k}$-points) and the nature of the surface material (metals need more $\mathbf{k}$-points than semiconductors or insulators).

Strengths and weaknesses

The ubiquity of DFT within computational surface science reflects the success it has enjoyed in describing surface structure, including relaxation, reconstruction, and adsorption geometries. Furthermore, it has generally proved reliable in assessing the relative energies involved in these situations, although it should be stressed that energy comparisons involving the gas phase ought not to be uncritically believed. For example, whilst DFT will often correctly identify the minimum-energy adsorption site for a given molecule on a particular surface, and perhaps even provide a reasonable estimate of the energy difference between this site and its competitors, calculated adsorption heats typically turn out to be quite significantly erroneous when compared with the most credible experimental measurements. Similarly, whilst activation energies for processes that take place entirely on the surface are often obtained with some accuracy, those relating to adsorption and desorption are much less reliable.

One reason for these specific failings is that the exchange-correlation functionals most suited for surfaces do not necessarily work well for gas-phase species (and vice versa) but estimates of adsorption energy require calculation of both adsorbed and isolated species. In the case of physisorbed species, moreover, further error arises because standard DFT is inherently incapable of capturing long-range dispersive interactions that involve excited-state contributions. Within regions of relatively uniform electron density, these interactions may be adequately mimicked by a mean-field exchange-correlation functional, but when regions of relatively high electron density are separated by an intervening region of low electron density this mimicry breaks down. A major trend in surface DFT studies, over the last decade or so, has been the emergence of various post hoc correction schemes, of greater or lesser sophistication and cost, that seek to account for these dispersive interactions. This trend continues rapidly to evolve, and the interested reader is referred to an in-depth review by Stöhr et al. (2019) for further details.

5.10 Exercises

5.1 The Ru{0001} surface has a structure in which hexagonal close-packed planes of atoms alternate in the ... *ABABAB* ... stacking sequence. Show explicitly that the LEED pattern in Fig. 5.2 is consistent with a p(2×2) overlayer, and explain why it displays six-fold rotational symmetry (in both spot position and intensity) when an idealised instance of the Ru{0001} surface conforms only to the *p3m1* space group.

5.2 An ARUPS experiment is performed on a Cu{111} sample (work function 4.94 eV) using light from the He I emission line (58.43 nm). What is the maximum possible surface-parallel wavevector component, k_{xy}, that can be accessed when detecting electrons emitted from 1 eV below the Fermi level? How accurately must emission angles be measured if one wishes to achieve a resolution of 0.05 Å^{-1} in k_{xy} throughout its range for these electrons? And how would your answers change if light from the He II emission line (30.38 nm) were used instead?

5.3 Based solely upon the polarisation-dependence of the furan NEXAFS spectra depicted in Fig. 5.13, what can be said about the orientation of this molecule on the surface?

5.4 Explain why the shapes of the sticking probability and adsorption heat curves shown in Figs. 5.17 and 5.18 imply the interpretations (i.e. precursor-mediated adsorption, direct adsorption, oxidation) noted on those diagrams.

5.5 Molecular nitrogen adsorbs on a flat and featureless metal surface. Discuss whether one should expect to see absorption attributable to the N–N stretch mode in a RAIRS experiment, if either (i) the molecules bind to the surface end-on, via a single nitrogen atom each, or (ii) the molecules lie flat on the surface, binding through both of their nitrogen atoms.

5.11 **Summary**

- Ultra-high vacuum conditions are commonly employed, serving the twin functions of permitting preparation of clean surfaces whilst reducing attenuation of photon, electron, or molecular beams travelling to and from the sample.

- Surface symmetry and structure may readily be probed by diffraction techniques, using low-energy electrons at normal incidence.

- Scanning probe microscopies, based upon either tunnel current or atomic forces, may be used not only to image surfaces but also to measure local electronic, vibrational, elastic, or other properties.

- Characteristics of occupied electronic states may be measured by means of photoemission spectroscopy. Where a suitable light source is not available, Auger electron spectroscopy may be a convenient alternative.

- Analysis of near-edge fine structure in X-ray absorption spectra gives access to unoccupied electronic states, and dependence on polarisation provides important clues as to adsorbate orientation.

- Supersonic molecular beams can be manipulated to vary the energy content of species impinging upon a surface, permitting systematic study of adsorption and reaction kinetics/thermodynamics.

- Various infrared absorption techniques can be used to confirm the identity of adsorbates, and together with the surface selection rule may also provide information on adsorbate orientation.

- Density functional theory is capable of describing many key properties of surface physics and chemistry, often assisting in the interpretation of experimental results.

Further reading

Arble, C., Jia, M., and Newberg, J. T., 2018. 'Lab-Based Ambient Pressure X-Ray Photoelectron Spectroscopy from Past to Present'. *Surf. Sci. Rep.* 73 (2): 37.

Baber, A. E., Tierney, H. L., and Sykes, E. C. H., 2010. 'Atomic-Scale Geometry and Electronic Structure of Catalytically Important Pd/Au Alloys'. *ACS Nano* 4 (3): 1637.

Baraldi, B., Lizzit, S., and Paolucci, G., 2000. 'Identification of Atomic Adsorption Site by means of High-Resolution Photoemission Surface Core-Level Shift: Oxygen on Ru(10$\bar{1}$0)'. *Surf. Sci.* 457 (1–2): L354.

Duflot, D., Flament, J.-P., Giuliani, A., Heinesch, J., and Hubin-Franskin, M. J., 2003. 'Core Shell Excitation of Furan at the O 1s and C 1s Edges: An Experimental and *Ab Initio* Study'. *J. Chem. Phys.* 119 (17): 8946.

Hohenberg, P., and Kohn, W., 1964. 'Inhomogeneous Electron Gas'. *Phys. Rev.* 136 (3B): B864.

Jenkins, S. J., 2020. 'Crystallography of Surfaces'. In *Springer Handbook of Surface Science*, edited by Rocca, M., Rahman, T. S., and Vattuone, L. Berlin: Springer International Publishing.

Kim, B., Kim, C. H., Kim, P., Jung W., Kim, Y., Koh, Y., Arita, M., Shimada, K., Namatame, H., Taniguchi, M., Yu, J., and Kim, C., 2012. 'Spin and Orbital Angular Momentum Structure of Cu(111) and Au(111) Surface States'. *Phys. Rev. B* 85 (19): 195402.

King, D. A., and Wells, M. G., 1972. 'Molecular Beam Investigation of Adsorption Kinetics on Bulk Metal Targets: Nitrogen on Tungsten'. *Surf. Sci.* 29 (2): 454.

Knight, M. J., Allegretti, F., Kröger, E. A., Polcik, M., Lamont, C. L. A., and Woodruff, D. P., 2008. 'The Adsorption Structure of Furan on Pd(111)'. *Surf. Sci.* 602 (14): 2524.

Kohn, W., and Sham, L. J., 1965. 'Self-Consistent Equations Including Exchange and Correlation Effects'. *Phys. Rev.* 140 (4A): A1133.

Liao, K., Fiorin, V., Gunn, D. S. D., Jenkins, S. J., and King, D. A., 2013. 'Single-Crystal Adsorption Calorimetry and Density Functional Theory of CO Chemisorption on *fcc* Co{110}'. *Phys. Chem. Chem. Phys.* 15 (11): 4059.

Liao, K., Fiorin, V., Jenkins, S. J., and King, D. A., 2012. 'Microcalorimetry of Oxygen Adsorption on *fcc* Co{110}'. *Phys. Chem. Chem. Phys.* 14 (20): 7528.

Mönig, H., Amirjalayer, S., Timmer, A., Hu, Z., Liu, L., Díaz Arado, O., Cnudde, M., Strassert, C. A., Ji, W., Rohlfing, M., and Fuchs, H., 2018. 'Quantitative Assessment of Intermolecular Interactions by Atomic Force Microscopy Imaging Using Copper Oxide Tips'. *Nat. Nanotech.* 13 (5): 371.

Narloch, B., Held, G., and Menzel, D., 1995. 'A LEED-*I/V* Determination of the Ru(001)-p(2×2)(O+CO) Structure: A Coadsorbate-Induced Molecular Tilt'. *Surf. Sci.* 340 (1–2): 159.

Salmeron, M., and Schlögl, R., 2008. 'Ambient Pressure Photoelectron Spectroscopy: A New Tool for Surface Science and Nanotechnology'. *Surf. Sci. Rep.* 63 (2): 169.

Sholl, D. S., and Steckel, J. A., 2009. *Density Functional Theory: A Practical Introduction* Hoboken: John Wiley & Sons.

Stöhr, M., Van Voorhis, T., and Tkatchenko, A., 2019. 'Theory and Practice of Modeling van der Waals Interactions in Electronic-Structure Calculations'. *Chem. Soc. Rev.* 48 (15): 4118.

Temprano, I., Liu, T., and Jenkins, S. J., 2017. 'Activity of Iron Pyrite Towards Low-Temperature Ammonia Production'. *Catal. Today* 286 (1): 101.

Woodruff, D. P., and Delchar, T., 1994. *Modern Techniques of Surface Science*. Cambridge: Cambridge University Press.

Glossary

1BZ See **first Brillouin zone**.

Absolute Coverage The concentration of adsorbates at a surface, expressed in units independent of the system described (e.g. molecules.m^{-2}) (see **coverage** and **relative coverage**).

Acoustic Describes phonons in which atoms within a single primitive unit cell move in phase (see **optical**, **phonon**, **primitive unit cell**).

Adsorbate An atom or molecule that is bound to a surface (see **substrate**).

Adsorption The arrival at, and binding of a species to, the surface of a material (see **desorption**).

Advancing Contact Angle Contact angle measured for a droplet that is increasing in size (see **contact angle** and **receding contact angle**).

AES Auger Electron Spectroscopy.

AFM Atomic Force Microscopy.

AFM-IR AFM InfraRed spectroscopy (see **AFM**).

ARUPS Angle-Resolved UPS (see **UPS**).

Atop Site Binding site in which an adsorbate sits directly above a single substrate atom (see **adsorbate**, **bridge site**, **hollow site**, **substrate**).

ATR Attenuated Total Reflectance.

Auger Electron An electron emitted as the result of energy transferred by refilling of a core hole (see **core hole**, **down electron**).

Back-Donation Refers to transfer of electrons from the substrate to an adsorbate (see **adsorbate**, **donation**, **substrate**).

BEP See **Brønsted–Evans–Polanyi relation**.

BET See **Brunauer–Emmett–Teller isotherm**.

Bravais Lattice A set of points defined by integer linear combination of a set of two or three (depending on dimensionality) defining vectors (see **real-space lattice** and **reciprocal lattice**).

Bridge Site Binding site in which an adsorbate bridges between two substrate atoms (see **adsorbate**, **atop site**, **hollow site**, **substrate**).

Brønsted–Evans–Polanyi Relation An empirical link between a change in reaction enthalpy and a corresponding change in activation barrier when considering several similar reactions.

Brunauer-Emmett-Teller Isotherm A relation predicting coverage as a function of gas-phase pressure, allowing for the possibility of multilayer formation (see **coverage** and **multilayer**).

Bulk Potential Energy The potential energy of an electron far below a given surface (see **vacuum potential energy**).

Capillary A narrow tube within which an exposed liquid surface is likely to be strongly curved, leading to significant effective forces.

Chemisorption The formation of a strong adsorbate–substrate bond due to chemical interactions (see **adsorbate**, **physisorption**, **substrate**).

Clausius–Clapeyron Equation A thermodynamic relationship between temperature and pressure that holds so long as two phases maintain equilibrium (i.e. no net shift from one phase to the other).

Complete Dewetting The condition when a droplet's contact angle is 180° (see **contact angle**).

Complete Wetting The condition when a droplet's contact angle is zero (see **contact angle**).

Contact Angle The angle between the surface of a liquid droplet or meniscus and the surface with which it makes contact.

Contact Mode A mode of AFM operation in which the tip experiences only repulsive forces (see **AFM**, **non-contact mode**, **tapping mode**).

Conventional Unit Cell A unit cell larger than the primitive unit cell, usually employed because it displays more symmetry than the smaller alternative (see **primitive unit cell**, **unit cell**).

Core Hole A shortfall in the complete filling of atomic core states due to emission of an electron.

Coverage The concentration of adsorbates at a surface. When not qualified by the adjectives 'absolute' or 'fractional/relative' the coverage will often be expressed in monolayer units (see **absolute coverage**, **monolayer**, **relative coverage**).

Cracking Pattern The characteristic distribution of fragments recorded by a mass spectrometer upon exposure to a given molecular species; it includes masses smaller than that of the parent species, arising due to ionisation-induced dissociation.

Critical Micelle Concentration The concentration beyond which micelles will form in preference to any increase in a given surfactant's surface excess (see **micelle**, **surface excess**, **surfactants**).

Critical Surface Tension The surface tension at which the contact angle of a liquid droplet on a given surface just vanishes (see **contact angle**, **surface tension**).

Desorption The unbinding/departure of an adsorbed species from a surface (see **adsorption**).

DFT Density Functional Theory.

Dispersion The variation of an electronic or vibrational band's energy as a function of wavevector.

Dividing Plane The location of the surface, for thermodynamic purposes, this plane notionally divides perfect bulk from perfect vacuum.

Donation Refers to transfer of electrons from an adsorbate to the substrate (see **adsorbate**, **back-donation**, **substrate**).

Down Electron An electron that refills a core hole from a higher-energy state.

EELS Electron Energy Loss Spectroscopy (see **HREELS**).

Effusive Molecular Beam A collimated stream of gas whose constituent particles conform closely to predictions from kinetic theory (see **supersonic molecular beam**).

Eley–Rideal Mechanism A model for surface reactions in which a gas-phase species directly impinges upon an adsorbed species (see **Langmuir–Hinshelwood mechanism**).

EXAFS Extended X-ray Absorption Fine Structure (see **SEXAFS**).

Final-State Effect A shift in the apparent binding energy of core levels in XPS due to the local relaxation of valence electrons in consequence of the photoemission process (see **initial-state effect**, **XPS**).

First Brillouin Zone The region of the reciprocal lattice lying closer to the origin than to any other reciprocal lattice point (see **1BZ**, **reciprocal lattice**).

First-Order Desorption Desorption occurring with a rate proportional to the coverage (see **coverage**, **second-order desorption**, **zero-order desorption**).

Flat Description of a surface with no atomic steps or kink atoms (see **kink**, **step**).

Fractional Coverage Synonym for relative coverage.

Friedel Oscillations Periodic decaying undulations in valence electron density due to scattering from a defect of some sort (e.g. from the surface of a bulk material).

Frustrated Rotations Vibrational motions of an adsorbate that would correspond to rotational motion were it not for restoring forces due to interaction with the substrate (see **frustrated translations**).

Frustrated Translations Vibrational motions of an adsorbate that would correspond to translational motion were it not for restoring forces due to interaction with the substrate (see **frustrated rotations**).

FTIR Fourier Transform InfraRed spectrometry.

Fuchs–Kliewer Modes Surface optical phonons found at frequencies above the highest bulk modes (see **optical**, **phonon**).

GGA Generalised Gradient Approximation.

Gibbs Isotherm A relation between the surface excess of different species and the rate of change in specific surface energy as a function of their chemical potential (or bulk activity/concentration) (see **specific surface energy**, **surface excess**).

GIXD Grazing Incidence X-ray Diffraction.

Grand Potential A thermodynamic potential quantifying a system's ability to do reversible work under conditions of constant temperature and constant chemical potentials (see **thermodynamic potential**).

Heat of Adsorption The heat released when a species adsorbs on a surface (see **adsorption**).

Helmholtz Energy A thermodynamic potential quantifying a system's ability to do reversible work under conditions of constant temperature and constant particle quantities (see **thermodynamic potential**).

Hermann–Mauguin Notation A descriptive approach to the labelling of space groups (see **space group**).

High-Index Surfaces Surfaces with one of more Miller indices of magnitude (much) greater than unity (see **Miller indices**).

Hollow Site Binding site in which an adsorbate nestles in the hollow created by three or four substrate atoms (see **adsorbate**, **atop site**, **bridge site**, **substrate**).

HOMO Highest Occupied Molecular Orbital (see **LUMO**).

HREELS High-Resolution Electron Energy Loss Spectroscopy (see **EELS**).

Ideal Surface An unachievable version of a surface in which all the atoms that are present maintain the exact positions they would have occupied if the bulk material were not truncated.

IETS Inelastic Electron Tunnelling Spectroscopy.

Initial-State Effect A shift in the apparent binding energy of core levels in XPS due to the different electrostatic potentials experienced by different atoms (see **final-state effect**, **XPS**).

Internal Energy A thermodynamic potential quantifying the total kinetic and potential energies of a system's constituent particles, excluding contributions due to the wholesale motion or position of the system itself (see **thermodynamic potential**).

Isostere A dataset (e.g. pressure versus temperature) obtained at constant coverage (see **isotherm**).

Isotherm A dataset (e.g. coverage versus pressure) obtained at constant temperature (see **isostere**).

Jahn–Teller distortion A reconstruction in which point symmetry is spontaneously reduced (see **Peierls distortion**, **reconstruction**).

Jellium A simplified model of the solid state, in which the positive charge of the ion cores is considered to be uniform (as will be the negative charge of the valence electrons, if unperturbed).

Kelvin Equation A relation describing changes in saturated vapour pressure above a surface, as a function of the surface curvature (see **saturated vapour pressure**).

King and Wells Technique A method for measuring sticking probability in supersonic molecular beam experiments (see **sticking probability**, **supersonic molecular beam**).

Kink A low-coordination site where an otherwise straight step features a localised zig-zag (see **step**, **terrace**).

Kisliuk Isotherm A relation predicting relative surface coverage in a similar way to the Langmuir isotherm, but accounting for the possibility of precursor states (see **coverage**, **Langmuir isotherm**).

KPFM Kelvin Probe Force Microscopy.

Langmuir–Hinshelwood Mechanism A model for surface reactions in which the involved species are adsorbed before the key reaction step occurs (see **Eley–Rideal mechanism**).

Langmuir Isotherm A relation predicting relative coverage at a solid surface as a function of the gas-phase pressure of an adsorbing species, subject to several simplifying assumptions (see **relative coverage**).

LDA Local Density Approximation.

LEED Low-Energy Electron Diffraction.

Longitudinal Describes phonons in which atomic motion occurs only parallel to the wavevector (see **phonon, transverse**).

Love Modes Surface acoustic phonons of shear horizontal polarisation (see **acoustic, phonon, Rayleigh modes, shear horizontal**).

LUMO Lowest Unoccupied Molecular Orbital (see **HOMO**).

Matrix Notation A general method of recording changes in surface periodicity due to either adsorbed overlayers or reconstruction (see **overlayer**, **reconstruction**, **Wood's notation**).

MFM Magnetic Force Microscopy.

Micelle A closed shell formed by surfactants within the bulk of a host liquid (see **surfactants**).

Miller Indices A trio of integers that together specify the crystallographic orientation of a surface.

ML MonoLayer (see **monolayer**).

Monolayer A unit of coverage, corresponding to one adsorbate per primitive unit cell of the ideal surface's real-space lattice (see **coverage**, **primitive unit cell**).

Motif A set of atoms associated with each point within a real-space lattice (see **real-space lattice**).

Multilayer An overlayer in which adsorbates build on underlying adsorbates to create a film comprising multiple layers (see **adsorbate**, **overlayer**).

NEXAFS Near-Edge X-ray Absorption Fine Structure (see **XANES**).

Non-Contact Mode A mode of AFM operation in which the tip experiences only attractive forces (see **AFM**, **contact mode**, **tapping mode**).

Normal Strain Strain components relevant to changes in area or volume (see **shear strain**).

Normal Stress Stress components relevant to changes in area or volume (see **shear stress**).

Optical Describes phonons in which some atoms within the primitive unit cell move in antiphase (see **acoustic**, **phonon**, **primitive unit cell**).

Overlayer A film of adsorbates on a substrate (see **adsorbate**, **substrate**).

Passivation The removal of Tamm states by saturation of dangling bonds upon adsorption (see **adsorption**, **Tamm states**).

Peak-to-Peak Height A proxy for intensity when analysing sharp features in a derivative spectrum.

Peierls Distortion A reconstruction in which translational symmetry is spontaneously reduced (see **Jahn–Teller distortion**, **reconstruction**).

PES PhotoEmission (or PhotoElectron) Spectroscopy (see **photoemission**, **UPS**, **XPS**).

Phonon A quantised vibrational excitation in a liquid or solid material.

Photoemission The ejection of an electron from a material upon absorption of light.

Physisorption The formation of a weak adsorbate-substrate bond due to physical interactions (see **adsorbate**, **chemisorption**, **substrate**).

Point Group Synonym for Point Group of the Space Group.

Point Group of the Space Group (or Point Group) The symmetry group (often referred to simply as the point group) formed by replacing the space group's glide operations with reflection operations; screw operations with rotation operations; and discarding translation operations altogether.

Polanyi–Wigner Equation A relation predicting desorption rate based upon relative coverage, surface temperature, the desorption barrier, and a pre-exponential factor (see **desorption**, **relative coverage**).

Precursor State A weakly bound transient state into which an adsorbate may briefly adsorb prior to either desorption or permanent adsorption (see **adsorption**, **desorption**).

Primary Electron An electron ejected directly from a material following either absorption of a photon or collision with an incoming electron (see **secondary electron**).

Primitive Real-Space Lattice Vectors A set of three real-space lattice vectors spanning a primitive unit cell (see **primitive unit cell**, **real-space lattice vectors**).

Primitive Reciprocal Lattice Vectors A set of three reciprocal lattice vectors spanning a reciprocal volume equal to that of the 1BZ (see **1BZ**, **reciprocal lattice vectors**).

Primitive Unit Cell The smallest possible unit cell of a given crystal (see **unit cell**).

Principal Stress Axes Eigenvectors of the stress tensor (see **principal stress components**, **surface stress**).

Principal Stress Components Eigenvalues of the stress tensor (see **principal stress axes**, **surface stress**).

Projected Bulk Band Structure The projection of electronic or vibrational bands from the three-dimensional 1BZ of a bulk material onto the two-dimensional 1BZ of its surface (see **1BZ**).

RAIRS Reflection Absorption InfraRed Spectroscopy.

Rayleigh Modes Surface acoustic phonons polarised in the sagittal plane (see **acoustic**, **Love modes**, **phonon**, **sagittal plane**).

Reaction Coordinate The generalised coordinate by which one measures progress of a system, through its potential-energy landscape, along the minimum-energy path between two neighbouring local minima.

Real Space Three-dimensional space suitable for describing the positions of point-like particles (see **reciprocal space**).

Real-Space Lattice A real-space Bravais lattice replicating the periodicity of a crystal (see **Bravais lattice**, **real space**).

Real-Space Lattice Vectors A set of vectors defining the points of a real-space lattice relative to some origin (see **real-space lattice**).

Receding Contact Angle Contact angle measured for a droplet that is decreasing in size (see **advancing contact angle**, **contact angle**).

Reciprocal Lattice A reciprocal-space Bravais lattice useful for the analysis of wave-like excitations within a crystal (see **Bravais lattice**, **reciprocal space**).

Reciprocal Lattice Vectors A set of vectors defining the points of a reciprocal lattice relative to the origin, corresponding to the wavevectors of excitations commensurate with the corresponding real-space lattice (see **real-space lattice**, **reciprocal lattice**).

Reciprocal Space Three-dimensional space suitable for describing the wavevectors of wave-like excitations (see **real space**).

Reconstruction The displacement of near-surface atoms from their ideal bulk positions in a manner that lowers the surface symmetry (see **ideal surface**, **relaxation**).

Redhead Formula A relation providing an estimate of the desorption barrier, based upon knowledge of the temperature at which desorption is maximised for a given heating rate (see **desorption**).

Relative (or Fractional) Coverage The surface concentration of an adsorbed species, expressed as a fraction of its maximum achievable concentration in a single layer on the given surface (see **absolute coverage**).

Relaxation The displacement of near-surface atoms from their ideal bulk positions, without any lowering of surface symmetry (see **ideal surface**, **reconstruction**).

Ripening An increase in mean droplet (or cluster) size driven by evaporation from smaller ones and condensation onto larger ones (see **sintering**).

Sabatier Principle The assertion that activity is highest when reactant-catalyst binding is of intermediate strength (see **volcano plot**).

Sagittal Plane Used to describe a phonon in which atoms are displaced only within the plane containing both the wavevector and the surface normal vector (see **phonon**, **shear horizontal**).

Saturated Vapour Pressure The pressure at which the gas and liquid phases of a substance can coexist at equilibrium (given other conditions, such as temperature or surface curvature).

SCAC Single-Crystal Adsorption Calorimetry.

Secondary Electron An electron ejected from a material due to collisions following from the excitation of a different electron (see **primary electron**).

Second-Order Desorption Desorption occurring with a rate proportional to the square of the coverage (see **coverage**, **first-order desorption**, **zero-order desorption**).

SEXAFS Surface Extended X-ray Absorption Fine Structure (see **EXAFS**).

SFG Sum Frequency Generation.

Shear Horizontal Used to describe a phonon in which atoms exhibit displacements perpendicular to the sagittal plane (see **phonon**, **sagittal plane**).

Shear Strain Strain components relevant to area- or volume-preserving deformation (see **normal strain**).

Shear Stress Stress components relevant to area- or volume-preserving deformation (see **normal stress**).

Shockley States Nearly-free-electron-like surface states, characterised by parabolic dispersion (see **dispersion**, **surface states**).

Shuttleworth Equation A relation linking surface stress and surface strain (see **surface strain**, **surface stress**).

Sintering An increase in mean droplet (or cluster) size driven by coalescence (see **ripening**).

Softening A drop in phonon frequency, possibly indicating the onset of a phase transition (see **phonon**).

Space Group A symmetry group comprising the complete set of symmetry operations pertaining to an extended object, including glide, screw, and translation (see **point group of the space group**).

Specific Surface Energy The reversible work done in creating unit area of fresh surface by cleaving a bulk sample.

Square, Triangular, Rectangular, Rhombic, Oblique The only five two-dimensional Bravais lattices (see **Bravais lattice**).

Steering The phenomenon whereby an incoming adsorbate is guided by the potential-energy landscape through the adsorption transition state (see **trapping**).

Step A linear boundary between two terraces of different elevation (see **kink**, **terrace**).

Sticking Probability The probability that a would-be adsorbate impinging upon a surface will, in fact, adsorb (see **adsorbate**, **adsorption**).

STM Scanning Tunnelling Microscopy.

STS Scanning Tunnelling Spectroscopy.

Substrate A surface upon which adsorbates are bound (see **adsorbate**).

Supersonic Molecular Beam A collimated stream of gas in which the translational motion of particles is nearly monoenergetic, while rotational and vibrational motion is strongly quenched (see **effusive molecular beam**).

Surface Active Agents Species that accumulate at the surface of a liquid and reduce its specific surface energy (see **specific surface energy**).

Surface Dipole The electrostatic dipole due to charge distribution at the surface.

Surface Excess Material not accounted for by a model of the surface in which particle density drops abruptly from its bulk value to zero in crossing the dividing plane (see **dividing plane**).

Surface Excess Entropy That portion of a system's entropy associated with its surface excess (see **surface excess**).

Surface Excess Grand Potential That portion of a system's grand potential associated with its surface excess and with the reversible work of surface creation by cleavage from a notional parent bulk material (see **dividing plane**).

Surface Excess Helmholtz Energy That portion of a system's Helmholtz energy associated with its surface excess and with the reversible work of surface creation by cleavage from a notional parent bulk material (see **dividing plane**).

Surface Excess Internal Energy That portion of a system's internal energy associated with its surface excess and with the reversible work of surface creation by cleavage from a notional parent bulk material (see **dividing plane**).

Surface Free Energy Usually a synonym for surface excess grand potential.

Surface Resonances Surface states that cross into allowed regions of the projected bulk band structure and are thus only poorly surface-localised (see **projected bulk band structure**, **surface states**).

Surface Selection Rule The constraint that infrared detection of vibrational modes is limited to those associated with a dynamic dipole moment perpendicular to the surface (see **RAIRS**).

Surface States Electronic or phononic states that are strongly localised at the outermost layers of the material, existing outside the allowed regions of the projected bulk band structure (see **projected bulk band structure**, **surface resonances**).

Surface Strain A tensor that quantifies the deformation of a solid surface (see **surface stress**).

Surface Stress A tensor that quantifies the forces opposing surface strain (see **surface strain**).

Surface Tension That property of a liquid surface whereby it tends to resist expansion of the surface area (see **surface stress**).

Surfactants Synonym for surface active agents.

Tamm states Dangling bond states, often characterised by rather flat dispersion (see **dispersion, surface states**).

Tapping Mode A mode of AFM operation in which the tip alternately experiences attractive and repulsive forces (see **AFM**, **contact mode**, **non-contact mode**).

Terrace An extended region of locally flat surface, bounded by steps (see **kink**, **step**).

Thermodynamic Potential Any one of a number of variables that in some manner quantify the state of a thermodynamic system (see **grand potential**, **Helmholtz energy**, **internal energy**).

Total Electron Yield A convenient proxy for the X-ray absorption coefficient of a sample.

TPD Temperature-Programmed Desorption.

Transverse Describes phonons in which atomic motion occurs only orthogonal to the wavevector (see **longitudinal, phonon**).

Trapping The phenomenon whereby an incoming adsorbate enters a precursor state and thus may make many attempts to cross the transition state to permanent adsorption (see **precursor**, **steering**).

Unit Cell A subset of a crystal structure, spanned by real-space lattice vectors, that may be tessellated to reproduce the entire crystal structure (see **real-space lattice vectors**).

UPS Ultraviolet Photoemission (or Photoelectron) Spectroscopy (see **ARUPS, PES**).

Vacuum Potential Energy The potential energy of an electron far outside a given surface (see **bulk potential energy**).

Vicinal Surface A surface that is slightly misoriented relative to a certain flat surface (see **flat**).

Volcano Plot A graph of reaction rate versus some key parameter of different possible catalysts, typically showing a peak when two competing rate-determining steps are balanced (see **Sabatier principle**).

Wood's Notation A widely used method of recording changes in surface periodicity due to either adsorbed overlayers or reconstruction (see **matrix notation**, **overlayer**, **reconstruction**).

Work Function The minimum energy that must be supplied to eject an electron in a photoemission experiment (see **bulk potential energy**, **photoemission**, **vacuum potential energy**).

XANES X-ray Absorption Near-Edge Spectroscopy (see **NEXAFS**).

XPS X-ray Photoemission (or Photoelectron) Spectroscopy (see **PES**).

XRD X-ray Diffraction.

Young's Equation A relation linking the contact angle of a liquid droplet with its surface tension and substrate-dependent stresses (see **contact angle**, **surface tension**).

Young–Laplace Equation A relation linking the surface tension and radius of a curved surface with the pressure differential between its concave and convex sides (see **surface tension**).

Zero-Order Desorption Desorption occurring with a rate independent of coverage (see **coverage, first-order desorption, second-order desorption**).

Zisman Plot A graph of the contact-angle cosine versus surface tension for a series of different liquids on a given surface (see **contact angle, critical surface tension, surface tension**).

Index

1BZ, *see* first Brillouin zone

absolute coverage 37, 111; *see also* relative coverage

adatoms 39, 40, 46, 60, 66

adsorption:
  activated 74–8, 92, 112
  activation barrier for 72, 74–8, 81, 91, 119
  chemisorption 64, 65–8, 70, 77
  enhanced by translational or vibrational energy 75–6
  enhanced by steering or trapping 76–8, 92
  equilibrium with desorption 18–21, 22–3
  forming ordered overlayer, *see* superstructure
  heat of 1, 18, 21, 22–5, 26, 73–4, 81, 82, 85, 89, 90, 111–2, 119, 120
  non-activated 72–4, 91, 112
  physisorption 64–5, 70, 77, 101–2, 120
  rate of 17–21, 71–2, 91, 95
  sites, *see* surface sites

advancing contact angle, *see* contact angle

AES, *see* Auger Electron Spectroscopy

AFM, *see* Atomic Force Microscopy

Anderson localisation 85–6

Angle-Resolved Ultraviolet Photoemission Spectroscopy (ARUPS) 60, 103–4, 120; *see also* UPS

annealing 94–5

ARUPS, *see* Angle-Resolved Ultraviolet Photoemission Spectroscopy

Atomic Force Microscopy (AFM) 98, 100–2, 121

atop site, *see* surface sites

ATR, *see* Attenuated Total Reflectance

Attenuated Total Reflectance (ATR) 113–4

Auger electron 105–6, 107

Auger Electron Spectroscopy (AES) 105–6, 121

back-donation 67, 85

BEP relation, *see* Brønsted-Evans-Polanyi relation

BET isotherm, *see* Brunauer-Emmett-Teller isotherm

bias voltage 98–100

binding energy (of electrons) 102–3, 104, 106, 107

Bloch's theorem 56–7, 82, 96, 117

Blyholder's covalent model of chemisorption 67

Bravais lattice 27–8, 31–2; *see also* real-space lattice

bridge site, *see* surface sites

Brønsted-Evans-Polanyi (BEP) relation 87–9, 92

Brunauer-Emmett-Teller (BET) isotherm 19, 21–2, 26; *see also* isotherms

bulk potential energy 52–3, 54, 55, 98, 102

capillary action 11–2, 16, 25, 26

catalysis 14, 69, 89–90, 91, 92, 105

chemisorption, *see* adsorption

Clausius-Clapeyron equation 22–3, 25, 26

concentration 3, 14–6, 17, 26, 112

contact angle 9–10, 11–2, 25, 26

contact mode, *see* Atomic Force Microscopy

conventional unit cell, *see* unit cell

core electrons and/or states 48, 102, 104–8

core hole 105–6

correlation effects 54, 55, 64, 69, 116–7, 120; *see also* exchange-correlation potential

corrosion 23, 105

cracking pattern 78–9

critical micelle concentration 16, 26; *see also* surfactants

critical surface tension, *see* surface tension

crystal structure:
  body-centred cubic 30, 37
  caesium chloride 29
  diamond 29
  face-centred cubic 29, 33, 35, 36, 37–9, 40–1, 43, 46
  hexagonal close-packed 37, 120
  simple cubic 29, 31
  zincblende structure 29

dangling bonds 39–40, 60–3, 68, 70, 99; *see also* Tamm states

Density Functional Theory (DFT) 115–20, 121

desorption:
  activation barrier for 74, 79–82, 91, 92, 119
  equilibrium with adsorption 18–21, 22–3
  first-order 80, 81–2, 91, 92
  pre-exponential factor for 79–81, 91, 92
  rate of 17–21, 79–81, 89, 91, 92
  Redhead formula for 80–2, 91, 92
  second-order 80, 91
  zero-order 79–80, 91

DFT, *see* Density Functional Theory

dividing plane 1–3, 4, 5, 8, 51–2

donation 67

down electron 105–6

drain current 107

dynamic dipole moment 112–3

effusive molecular beam 109

elastic emission (of electrons) 103–5, 107

elastic scattering (of electrons), *see* electron diffraction

elastic tunnelling (of electrons) 98–100

electron diffraction 95–7; *see also* LEED

Eley-Rideal mechanism 90, 91

enthalpy:
  of adsorption 18; *see also* adsorption, heat of
  of reaction 87–9, 92

entropy:
  as factor in adsorption or reaction 18, 71–2, 76, 87, 90, 91
  bulk 4, 15
  surface excess 5

Ewald sphere 96

EXAFS, *see* Extended X-ray Absorption Fine Structure

exchange-correlation potential 55, 64, 116; *see also* exchange-correlation functional

exchange-correlation functional 116–7, 120

exchange effects 54, 55, 64, 69; *see also* exchange-correlation potential

Extended X-ray Absorption Fine Structure (EXAFS) 107

Fermi energy (or level) 54–5, 58–63, 65–8, 70, 98–100, 102–4, 107, 120

Fermi surface (or sphere) 50, 58

Fermi wavenumber 50–1, 52, 60

final-state effects 104

first Brillouin zone (1BZ):
  bulk 44, 57–9, 82–3
  surface 44–5, 46, 57–63, 68, 82–4, 96, 117, 119

Fourier Transform InfraRed (FTIR) 112

fractional coverage, *see* relative coverage

free-electron (or nearly-free-electron) behaviour 48–55, 56, 58, 59–60, 65, 67, 70, 103

Friedel oscillations 48–52, 60, 69

frustrated rotational or translational modes 73, 82, 86–7, 92, 113

FTIR, *see* Fourier Transform InfraRed

Fuchs-Kliewer modes 84; *see also* surface-localised phonon modes

Generalised Gradient Approximation (GGA) 117

GGA, *see* Generalised Gradient Approximation

Gibbs isotherm 14–6, 26; *see also* isotherms

Gibbs-Duhem equation 15

GIXD, see Grazing Incidence X-ray Diffraction
grand potential:
  bulk 4
  surface excess 4, 5, 6, 25, 33
Grazing Incidence X-ray Diffraction (GIXD) 95
Gurney's ionic model of chemisorption 65-6

Hartree potential 116
Helmholtz energy:
  bulk 4
  surface excess 4, 5, 7
Hermann-Maugin notation 31-3, 34
high-index surfaces 35, 46; see also vicinal
  surfaces
High-Resolution Electron Energy Loss
  Spectroscopy (HREELS) 113
Hohenberg-Kohn theorems 115, 116
hollow site, see surface sites
HREELS, see High-Resolution Electron Energy
  Loss Spectroscopy

ideal gas 13
ideal solution 16
ideal surface 30, 33, 37-42, 46, 62, 68
IETS, see Inelastic Electron Tunnelling
  Spectroscopy
Inelastic Electron Tunnelling Spectroscopy
  (IETS) 100
inelastic emission, scattering or tunnelling
  (of electrons) 96, 100, 103, 107
infrared selection rule 112; see also surface
  selection rule
initial-state effects 104
internal energy:
  bulk 4
  surface excess 4-5, 14-5
ion cores 48, 55
isosteres 22-3, 26, 112
isotherms 1, 22-3 , 24, 25, 36, 71, 78, 79
  Brunauer-Emmett-Teller, Kisliuk, and
    Langmuir isotherms 17-22, 26
  Gibbs isotherm 14-6, 26

Jahn-Teller distortion 61-3
jellium 48-52, 55, 56; see also free-electron
  (or nearly-free-electron) behaviour

Kelvin equation 13, 26
Kelvin Probe Force Microscopy (KPFM) 102
kinetic energy (of an adsorbate) 73-8, 90,
  91, 92
kinetic energy (of electrons) 52, 54, 102-4,
  106, 107, 116, 118, 119
King & Wells technique 110-1
kinks 33, 35, 46
Kisliuk isotherm 19, 20-1, 26; see also
  isotherms
Kohn-Sham equation 116-9
KPFM, see Kelvin Probe Force Microscopy

Langmuir isotherm 17-20, 21, 25, 26, 91;
  see also isotherms
Langmuir-Hinshelwood mechanism 90, 91
lateral interactions 1, 82
  attractive 24-5, 26
  repulsive 24, 26
lattice (2D):
  Bravais 31-2
  centered versus primitive 32-3
lattice (3D):
  Bravais 27-8
  body-centred cubic 29, 30
  face-centred cubic 29, 30, 44, 57
  orthorhombic 31
  simple cubic 28-9, 30
LDA, see Local Density Approximation
LEED, see Low-Energy Electron Diffraction
Local Density Approximation (LDA) 117
London dispersion 65
Love modes 84; see also surface-localised
  phonon modes
Low-Energy Electron Diffraction (LEED) 95-7,
  103, 120, 121

Magnetic Force Microscopy (MFM) 102
mass spectrometry 78-9
matrix notation 40, 42, 43, 46
MFM, see Magnetic Force Microscopy
micelles, see surfactants
Miller indices 30-1, 35, 37, 46
ML, see monolayer
molecular vibrational modes:
  gas-phase 73, 75-6, 110
  surface 82, 84-6, 87, 92, 100, 112-5
monolayer (ML) 36-7
motif 28-30, 31, 46

Near-Edge X-ray Absorption Fine Structure
  (NEXAFS) 107-8, 120, 121
NEXAFS, see Near-Edge X-ray Absorption Fine
  Structure
non-contact mode, see Atomic Force
  Microscopy
non-linear optics 114-5
normal strain 7
normal stress 8, 9

passivation 61-3
Pauli repulsion 65, 101
peak-to-peak height (in derivative
  spectra) 106
Peierls distortion 61-3
PES, see Photoemission (or Photoelectron)
  Spectroscopy
phonons 82-4, 85-6, 87, 92, 113
  longitudinal and transverse 83
  optical and acoustic 83-4
  sagittal plane and shear horizontal 83-4
  see also surface-localised phonon modes

photoemission 55-6, 60, 68, 102-5, 107, 121;
  see also UPS; XPS
Photoemission (or Photoelectron)
  Spectroscopy (PES); see UPS; XPS
physisorption, see adsorption
point group (of the space group) 28, 31
point symmetry 28-31, 37, 61
Polanyi-Wigner equation 79
precursor 20-1, 26, 71, 77, 111-2, 120
primary electron 105, 107
primitive real-space lattice vectors 27-31, 40,
  42, 44, 46; see also real-space lattice
primitive reciprocal lattice vectors 42-4, 46,
  63; see also reciprocal lattice
primitive unit cell, see unit cell
principal stress axes and components 9
projected bulk electronic band structure
  56-63, 68, 69-70
projected bulk vibrational band
  structure 82, 84
pseudopotentials 118

quantum corrals 60

RAIRS, see Reflection Absorption InfraRed
  Spectroscopy
Rayleigh modes 84; see also surface-localised
  phonon modes
reaction coordinate 74, 75, 76, 87, 90
real-space lattice:
  bulk 27-31, 57; see also lattice (3D)
  surface 27, 31-3, 40-2, 44, 46, 61-3, 85, 96;
    see also lattice (2D)
receding contact angle, see contact angle
reciprocal lattice:
  bulk 42-4, 47, 57, 117-9
  surface 27, 44-5, 46, 47, 57-8, 63, 96, 103
reconstruction 45, 46, 54, 71, 84, 96, 118, 119
  clock 38-9
  dimer 40, 61-3
  herringbone 98
  hex 37-8
  missing row 38-9
  see also superstructure
Reflection Absorption InfraRed Spectroscopy
  (RAIRS) 112-3, 114, 115, 120
relative coverage 17-25, 36-7, 63, 66, 71, 79-
  80, 92, 105; see also absolute coverage
relaxation 37-9, 46, 54, 69, 71, 119
R-factor 97
ripening 13-4, 16

sagittal plane 83-4, 113; see also phonons
salts (effect upon specific surface energy) 16
saturated vapour pressure 12-4, 21, 25, 26
SCAC, see Single-Crystal Adsorption Calorimetry
scanning probe techniques 97-102; see also
  AFM; IETS; KPFM; MFM; STM; STS
Scanning Tunnelling Microscopy (STM) 60,
  68, 98-100, 121

Scanning Tunnelling Spectroscopy (STS) 99–100
Schrödinger equation 59, 116
secondary emission (of electrons) 103, 106, 107
SEXAFS, *see* Surface Extended X-ray Absorption Fine Structure
SFG, *see* Sum-Frequency Generation
shear strain 7
shear stress 8, 9
Shockley states, *see* surface-localised electronic states
Shuttleworth equation 8, 26
Single-Crystal Adsorption Calorimetry (SCAC) 111–2
sintering 13–4
slab model 118–9; *see also* supercell
softening of phonon modes 84
space group 27, 28, 31–5, 37–9, 45, 46, 120
specific surface energy 4–5, 6, 7–8, 11, 14–6, 25–6, 33
sputtering 94–5, 106
steps 33, 35, 46, 60, 99
sticking probability 72, 76–8, 89, 91, 92, 110–1, 120
STM, *see* Scanning Tunnelling Microscopy
STS, *see* Scanning Tunnelling Spectroscopy
Sum-Frequency Generation (SFG) 114–5
supercell 118–9; *see also* slab model
supersonic molecular beam 76, 109–11, 121
superstructure 33, 40–2; *see also* matrix notation; Wood's notation
surface active agents, *see* surfactants
surface coverage, *see* absolute coverage; relative coverage
surface curvature 1, 11–4, 26
surface dipole 48, 52–5, 64–5, 66, 69
surface excess 2–3, 5, 15–6, 17, 26
surface excess entropy, *see* entropy
surface excess grand potential, *see* grand potential
surface excess Helmholtz energy, *see* Helmholtz energy
surface excess internal energy, *see* internal energy
Surface Extended X-ray Absorption Fine Structure (SEXAFS) 107
surface free energy 1, 5, 7, 25, 33, 39, 84; *see also* surface excess grand potential
surface-localised electronic states 48, 56–63, 70
    Shockley states 58, 59–60, 68, 70, 104
    Tamm states 60–3, 68–9, 70, 99

surface-localised phonon modes 82–4, 85, 86, 87, 92
    Fuchs-Kliewer, Love, and Rayleigh modes 84
surface reactions and reactivity 35, 69, 71, 87–90, 91, 92, 94–5, 109–10, 121
surface resonances 59, 60, 70; *see also* surface states
surface selection rule 113, 121; *see also* infrared selection rule
surface sites:
    atop 36, 46, 85
    bridge 36, 46, 85
    hollow 36, 39, 46, 85
surface states, *see* surface-localised electronic states; surface-localised phonon modes
surface strain 7–8
surface stress 1, 8–9, 25–6, 118
    measurement of 10
surface structure 27, 33, 35, 36, 46, 48, 71, 119, 121
surface symmetry 27, 31–4, 37, 46, 48, 61–3, 71, 121
surface tension 1, 5–7, 8, 11, 13, 14, 25–6
    critical 10
    measurement of 9–10
surfactants 1, 14, 16, 17, 26
symmetry operations (elements):
    glide (glide plane) 28, 32–3
    inversion (centre of inversion) 28, 114–5
    reflection (mirror plane) 28, 32–3
    rotation (rotational axis) 28, 32–3, 120
    screw (screw axis) 28
    translation (lattice vectors) 28, 37, 39, 40, 46, 61–3

Tamm states, *see* surface-localised electronic states
tapping mode, *see* Atomic Force Microscopy
Temperature-Programmed Desorption (TPD) 78–82, 91, 92
terraces 33, 35, 46
thermodynamic potentials 4–5, 25
thermodynamic processes:
    at constant chemical potential 4, 6, 7, 14, 15
    at constant particle quantity 4, 7
    at constant relative coverage 22–3
    at constant temperature 4, 6, 7, 14–6, 17–22
titration 94–5
total electron yield 107

TPD, *see* Temperature-Programmed Desorption
transition state 74–8, 87–9, 91, 117
tunnel current 98–100, 121

unit cell 68–9, 96
    centred 32, 41, 42
    conventional 29–30, 31, 46
    primitive 29–30, 31–2, 35, 36–7, 38, 40–3, 44, 45, 49, 57, 62, 83, 85
ultra-high vacuum 78, 94–5, 98, 102, 105, 108, 112–3, 121
Ultraviolet Photoemission Spectroscopy (UPS) 60, 102–4, 105
UPS, *see* Ultraviolet Photoemission Spectroscopy

vacancies 40, 68, 99
vacuum level (or potential energy) 52–6, 65–6, 67, 102, 105, 107
valence electrons and/or states 48–52, 54–5, 58, 65, 99, 102, 104, 118
vibrational spectroscopies 112–5; *see also* ATR; HREELS; RAIRS; SFG
vicinal surfaces 33, 35, 46; *see also* high-index surfaces
volcano plots 89–90, 92

wetting and dewetting 9
Wilhelmy plate 10
Wood's notation 40–1, 42, 46
work function 48, 54–6, 65–6, 68, 69, 118
    in AES 106
    in PES 102–3, 120
    in STM 98–9

XANES, *see* X-ray Absorption Near-Edge Structure
XPS, *see* X-ray Photoemission Spectroscopy
X-ray absorption coefficient 107–8
X-ray Absorption Near-Edge Structure (XANES), *see* NEXAFS
XRD, *see* X-Ray Diffraction
X-Ray Diffraction (XRD) 95
X-ray Photoemission Spectroscopy (XPS) 104–5, 107

Young's equation 10, 26
Young-Laplace equation 11, 12, 26

Zisman plot 10